Thomas Gray in Copenhagen

Thomas Gray in Copenhagen

In Which the Philosopher Cat Meets the Ghost of Hans Christian Andersen

Philip J. Davis

Illustrated by
Marguerite Dorian

COPERNICUS
AN IMPRINT OF SPRINGER-VERLAG

©1995 Springer-Verlag New York, Inc.
All rights reserved. No part of this publication may be reproduced,
stored in a retrieval system, or transmitted, in
any form or by any means, electronic, mechanical,
photocopying, recording, or otherwise, without the prior
written permission of the publisher.
All characters except for historical personages are fictitious.

Published in the United States by Copernicus, an imprint of
Springer-Verlag New York, Inc.

Copernicus
Springer-Verlag New York, Inc.
175 Fifth Avenue
New York, NY 10010
USA

Library of Congress Cataloging-in-Publication Data
Davis, Philip J., 1923–
    Thomas Gray in Copenhagen: in which the philosopher cat meets the
ghost of Hans Christian Andersen/by Philip J. Davis; illustrated
by Marguerite Dorian.
    p. cm.
    ISBN 0-387-94493-1 (alk. paper)
    1. Andersen, H.C. (Hans Christian), 1805–1875—Fiction.
2. Copenhagen (Denmark)—Fiction.   3. Cats—Fiction.   I. Title.
PS3554.A937653T46   1995
813'.54—dc20   95-13243

Manufactured in the United States of America.
Printed on acid-free paper.
Illustrated by Marguerite Dorian.
Designed by Irmgard Lochner.

9 8 7 6 5 4 3 2 1

ISBN 0-387-94493-1

To David and Isabelle Pingree

EJ BLOT TIL LYST*
Not only for pleasure

*Inscription on the proscenium arch
of the Royal Theater in Copenhagen

A rumor has spread in Copenhagen that the famous Danish writer of fairy tales, Hans Christian Andersen, did not have a high regard for cats. Thomas Gray, the Philosopher Cat from Pembroke College, and her colleague, the scholarly Lucas Fysst, determine to rehabilitate Andersen's reputation. This involves them in adventures in Copenhagen that provide readers with an enjoyable satire on academic thought as well as an introduction both to Andersen and to Denmark.

A part of the material of this book was filmed by Danish National Television for presentation on *Troldspejlet* (*The Magic Mirror*), a program for young teenagers.

Contents

	Foreword and Acknowledgments	xi
1	The Soul of a Canary	1
2	Lessons in the Danish Language and Literature	11
3	The Conference	19
4	The Search Begins	29
5	Conversation with a Bouillabaisse	35
6	Thomas Gray Arrives on the Scene	41
7	The Collaboration Renewed	47
8	Found on Stroeget	55
9	The Glory of the Realm	63
10	Spectral Dialogue in Nyhavn	67
11	Lucas Fysst's Dream: The Wienerbroed and the Cat	73
12	Instructions Are Received	81
13	The Plain Brown Envelope	91

14	The Kongekat's Story	101
15	Lunch at Villa Fredenskjold	117
16	Diamonds Around Her Neck	123
17	Tea at Amalienborg	131
18	Advanced Procedures	139
19	More Pressure Is Applied	147
20	"Perhaps YOU Have"	151
21	Den Lille Havfrue	159
22	Bona Fides Established	165
23	Chaos Off Camera	169
24	The Magic Mirror	177
25	A Good Bit More Is Revealed	183

Foreword

Nyhavn 18 is an eighteenth-century building situated centrally in a most picturesque part of Copenhagen, just off the harbor. In 1968, it was bought by Danmarks Nationalbank (which corresponds to the Bank of England or the Federal Reserve Bank in Washington, D.C.) and renovated for the purpose of providing flats for foreign scholars, scientists, artists, and writers for periods of up to a year.

A few years ago, I was giving a course of lectures at the Roskilde University Center, a half hour by train west of Copenhagen, and my wife and I were fortunate enough to be accommodated in one of these flats. Our pleasure was further enhanced when we learned that from 1873 until his death in 1875, Hans Christian Andersen, the famous writer of fairy tales, had occupied two of the rooms of our flat. This information is widely available. Tourists can find the address in the useful *Copenhagen This Week* pamphlet given out free of charge at every hotel. Danmarks Nationalbank had underlined this information for occupants of the flat by assembling and placing on its walls some Andersen memorabilia. Although we were living in a minimuseum, it was not open to the general public.

As I sat at a beautifully modern, carefully crafted desk of Danish design, I would look out the window to the other side of the canal and see the masts of the sailboats docked there. If I looked up at the wall in front of my desk, I would see a photograph of Hans Christian Andersen sitting in the very same position at his desk (of heavy mid-nineteenth century design), looking out the same window at substantially the same harbor scene. It was easy to believe that this room would be a very good place to write a story.

Some years before, I had written a satire of the Cambridge University dons. The story had a heroine, Thomas Gray (yes, a heroine), the very intelligent cat of Pembroke College. The hero was Dr. Lucas Fysst, a slightly eccentric classicist and historian of science. Together, Thomas and Lucas continued on a glorious career in the restricted circles of scholarly achievement.

The heroine of the story was based on a real cat named Thomas Gray who hung around Pembroke College. As a result of the story's publication and several months after its appearance, the real Thomas Gray, poised tensely in the hands of the Senior Tutor, received a two-page color spread in the magazine section of the *London Sunday Observer*. Mulling over this adventure, I thought: Thomas and Lucas deserve to be reunited in Nyhavn 18, where a master raconteur once lived. Why not?

Acknowledgments

I have always been amazed at the enormous number of credit lines that appear at the end of movies and TV videos. Although it is not customarily acknowledged, fiction, from conception through production, also merits a large number of credits. What follows is an abbreviated list for this book.

I should like to thank Bernhelm Booss-Bavnbek and Mogens Heefelt, who arranged for my lectures and for my stay in Copenhagen. I thank Ole Skovsmose for my stay in Skoerping and visit to the Troldskov. Thanks also to Marguerite Dorian, who started the story, and to Natalja Koot, who showed me how the story might continue. Simone Lang of Danmarks Radio, quite by accident, extended the story in a completely different direction. Elias Bredsdorff, an authority on Scandinavian literature and author of a most readable biography of H. C. Andersen, set me straight about a number of Andersen details and encouraged me to go forward. Edda Sveinsdottir led me to Faarevejle and Bothwell. Inger Soerensen corrected my misspellings of Danish words and reassured me in a number of ways. As always, David Pingree has been a fountain of lore. Last, but not least, I should like to thank Danmarks Nationalbank for providing the mise-en-scène.

I hope this story will arouse in my readers the desire to know more about Andersen. I should like to recommend two books:

Elias Bredsdorff, *Hans Christian Andersen: The Story of His Life and Work, 1805–1875,* New York, Charles Scribner's Sons, 1975.

Patricia L. Conroy and Sven H. Rossel (editors and translators), *The Diaries of Hans Christian Andersen,* Seattle, University of Washington Press, 1990.

P.J.D.

Note: Danish vowels not present in the English alphabet have been transliterated.

Thomas Gray in Copenhagen

1

The Soul of a Canary

Over apples and port, Ian Plumbley asked, "So. You're going to Copenhagen, are you, Fysst?" Plumbley was one of those people who show off by peeling an apple in one long continuous spiral.

"A short stay, really. Only several months. First a conference, you know. Then some work in the Kongelige Bibliotek, the Royal Danish Library. And then a few odd lectures here and there in Denmark, possibly in Germany as well. I've worked things out here so I can get away."

These words were spoken in the Combination Room of Pembroke College, Cambridge University, England. Ian Plumbley was the Stansfield Professor of Astrophysics. The second speaker was Dr. Lucas Fysst, Reader in the History of Sciences.

The supper had been a substantial one, featuring roast saddle of lamb, one of Plumbley's favourites, which he was now embellishing with glasses of dessert wine.

"And what, if I'm not being obtrusive, is the subject of your conference?"

"Not at all, not at all. It's a conference on the nature of the soul. A very significant topic for this generation, since we in the West—not I, mind you, and I'm of the West, nor you, I should hope—we of the West tend to be a godless, soulless, heroless crew."

Plumbley made a decided but slightly muted noise of disapprobation and put a piece of apple into his mouth with the spiral peel still dangling.

"The soul, ah yes, the soul. A metaphysical concept that won't go away, no matter how often we physicists point out that it plays no role whatever in mathematical theories of the universe."

"I should say then, Plumbley, that your theories are correspondingly limited. Probably outdated. Moreover, even in its current, limited state, you don't care to admit how much metaphysics there is built into your physics."

"Well, we'll not argue the point tonight. And what specifically will you talk about in Copenhagen?" Plumbley asked Lucas Fysst.

"I shall discuss whether animals have souls. I shall survey various Western opinions on the matter, beginning with the famous *De Anima* of Aristotle. I shall move to the later commentaries by Albertus Magnus, Thomas Aquinas, and John Philoponus. I shall end with a few slight thoughts of my own on the subject. The matter deepens considerably when one considers the opinions of the East, say, the opinions embodied in the *Chandogyopanisad* or in the *Bramasutras* of Bedareyana. But I shall not talk about that."

"Why do you avoid discussing Eastern opinion?" queried Professor Plumbley with mock irritation.

"Because there will be a greater authority at the conference."

"And who would that be?"

" Mahamahopadhyaye Jivatmanandasvemim. The Swami, for short."

"Yes, you'd be right in staying out of that man's way. But what are your personal thoughts on whether animals have souls?"

"My dear boy, please do come to Copenhagen and my opinions will be revealed in their entirety," Fysst offered.

"Yes, of course. Obtrusive of me, really. Do you know Copenhagen? Do you know the Danes?"

"Not at all."

"Well, I've been there many times," said Plumbley. "Great place for quantum physics, you know. Lovely people, Danes. Relaxed and all that. Lovely memories of Denmark."

"Tell me more. I'd find it useful. What's the weather like there? What should I take along?"

"You'll be there in February, March, and April? Then I should say the weather will be much like Cambridge, only a bit more disagreeable. Wetter, windier, colder. Copenhagen is a graveyard for umbrellas. Why don't you people meet in the British Virgin Islands? Now that's paradise on earth. The soul could enlarge itself there."

"So I'll need a scarf?"

"That you will. You'll find that the Danes are totally bundled up. I don't mean emotionally. Your typical Dane will twist a long and complicated scarf around the throat six or seven times before facing the elements. I fell in love once with a Danish lady astrophysicist. It was well into May before I could see she had a neck."

"A severe shock to you, I dare say. It's good, though, that from time to time faith is renewed by experience. Else it withers on the vine."

Professor Plumbley drained the bottom of his wine glass and rose to leave. "One more thing, Fysst. The Invigilation Committee may want to get in touch with you while you're away. They're thinking of putting in some changes in how exams are monitored. Could

I have your address?" He handed Lucas a pen and tore off a blank area of *The Sunday Independent*. Lucas wrote down:

L. Fysst
Nyhavn 18
1051 Koebenhavn K
Danmark

Lucas Fysst (his last name does *not* rhyme with "fist") was Reader in the History of Sciences, specializing in the sciences and lore of Rome, Greece, Egypt, and Mesopotamia. He had rooms in College but lived in Waterfen St. Willow with his wife, Barbara, and his young daughter, Cornelia. Though Lucas was an ordained minister of the Church of England and wore a clerical collar and shirt, he had no parish responsibilities. Whenever Lucas was approached by a stranger on the streets of Cambridge who said to him, "Padre, I have this problem . . . ," Lucas would invariably refer the stranger to St. Michael's Church on the Hills Road.

"I have no experience with the general sadnesses and tragedies of everyday life. I have my own, of course, but I find that my experience does not extend easily to others," Lucas would say. Perhaps the most recent personal sadness he was referring to was that he was badly outbid by a gentleman from Japan for a rare Paris 1694 edition of Ozanam's *Récréations mathématiques et philosophiques*. It was not a book he needed for his work, but he thought it would have been nice to own it.

One of Fysst's chance encounters with the woes of the public was to lead him to his Copenhagen conference. It happened that the previous April, late in the month, he was asked to conduct the

Evensong service in the Wren Chapel of Pembroke College. The service would normally have been taken by Dr. Montrose, the Dean of Pembroke, but Rosey, as he was called because of his apple cheeks, had to be in Lancashire to attend the wedding of his sister's daughter.

Be all this as it may, after the service, a woman who had been sitting at the back of the chapel and who was a total stranger to the

Pembroke milieu came up to Lucas Fysst and asked, "Doctor Montrose, will my canary go to heaven when he dies?"

This question was rather removed from Lucas's current line of thought; nonetheless, it required an answer, and he realized that no matter whether he answered yes or no, he would be in some kind of trouble.

"Dear lady, I don't know how Dr. Montrose would answer your question, and a very good question it is indeed. However, I shall consider the matter and let you have my opinion. Please do visit me in my rooms here in College in a few days."

The lady departed the scene with a light heart, believing that this good man would look into the individual merits of her particular bird. That was not at all what Fysst had in mind.

The scientific mind has its moments of generalization, and this was one of them. Lucas first transformed the problem of heaven into the problem of the immortality of the soul, a well-discussed question. From the specific canary, named Sweetie (let us say), resident in a brass cage at 2353 Milton Road (let us say), Lucas Fysst's mind wandered to canaries in general, from canaries to birds in general, and from birds to the animal kingdom—the fauna, so called. Arrived at that point, there was then no way for his mind to stop, and he considered the flora as candidates for souls. From the fauna and the flora it was an easy step to the whole of God's creation. And so, ultimately, the question in his mind became: Does the cosmos have a soul?

Once posed in this general way, Lucas Fysst's knowledge joined hands with his imagination. For Western opinion, he consulted Aristotle and other ancient authorities. He consulted the doctors of the early and mediaeval church. He searched the opinions of Paracelsus and the heretic Bruno. He peeked into the writings of Thomas Muggleton and the recent pronouncements of the latest three popes of the Roman Catholic Church, and he examined writings of the late Robert Ramsey, Archbishop of Canterbury, on the odd chance that any of those great leaders might have devoted some thought to the matter.

To all these opinions, which were variable in their profundity and authority, he added the opinions of the ancient ascetics and modern vegetarians and animal-rightists. The total picture was so confusing that it seemed to Lucas that nothing short of an international conference at the highest metaphysical level was required to untangle the jumble and to set humanity on the right track.

He mentioned the question to a number of friends on the Continent. Professor Mogens Hvidvin of Copenhagen came forward with a lecture room and a few odd kroner for basic expenses and in this way the conference was born.

While these matters were in motion, Lucas Fysst knew that he must also descend from the heights of philosophical speculation and deal with Sweetie of 2353 Milton Road (let us say). The unknown woman was certain to knock at his door, and he would have to bite the bullet. Knock she did, after a few days, and she received the following opinion:

"Dear lady, perhaps it was a serious omission on my part, but I have not been able to consider the specific merits of your canary. (Did you say her name was Sweetie?) Upon reflecting on your question, will your canary go to heaven, the answer depends upon whether your canary has a soul. And the answer to that question, within the bounds set out by recent British philosophy and metaphysics, would depend upon the interpretation one places on the word 'has.'

"But that, dear lady, lies entirely in the theoretical sphere of syntactics and semantics and possibly semiotics. Now you may ignore entirely what I have just said and consider my intuitive sensibility only: You have my personal assurance that your canary bird will go to heaven."

"You are a very kind and wise man, Dr. Montrose."

2

Lessons in the Danish Language and Literature

Lucas Fysst's travel plans were now set. He was to fly from London's Heathrow Airport to Hamburg, Germany. In Hamburg, he was to consult with Professor Dr. Dr. h.c. mult. Felix Topfenmeyer on a point in Heiberg's 1912 translation of *Hero of Alexandria*. The next day he was to meet with Stephanie Eriksdottir, a Copenhagen girl of fifteen, whose father some years back had worked with Lucas in the Vatican Library in Rome. Together they would go by train to Copenhagen.

This was Lucas's first trip ever to Copenhagen. Danish was not among the many languages he normally dealt with. But he argued that if he could spot the English, Scottish, Saxon, or German roots of any particular Danish word, he should be able to make do and get on in Copenhagen. In this respect, Lucas was not your typical Englishman who believes as firmly as two and two make twenty-two that English is the universal language of discourse and that one should therefore speak up publicly in English in a loud and commanding manner.

Seated in 17A, on the aisle of SAS Flight 306, Heathrow to Hamburg, Lucas found that 17B was occupied by an American who kept his ten-gallon Stetson hat on his head throughout the flight. Normally shy, normally burying his nose in the *Journal of Late Alexandrine Studies* or some such, Lucas felt rather forward that morning. His neighbor, a cheerful, pumpkin-faced man, opened the conversation.

"Howdy, friend. Pete Williams from Big Bend, Texas, the little old Chili Capital of the World. Chili con carne is what puts zing into your bing, if you know what I mean. On my way to Hamburg for a little ding ding with the folks at Deutsche Fabrikswerk, GmbH. I pronounce that GmbH word as 'gumbo.'"

Lucas determined to fight fire with fire and effusion with countergush. "Lucas F-y-s-s-t. Does not rhyme with 'fist' but with . . ." Lucas pointed heavenward, beyond the place in the plane where the breathing apparatus comes down automatically in the unlikely event of . . . Pete Williams nodded at the vertical gesture, and noting Lucas's ecclesiastical collar, he said, "I get it. Fysst. Rhymes with Christ." Then Lucas's neighbor raised and directed his own finger heavenward in a countergesture of total understanding.

Lucas continued, following his neighbor's model. "I'm from Cambridge. Final destination is Copenhagen, where I'm to lecture and to et cetera, if you get my meaning."

"You know," said Pete Williams, "I may just find myself in the Land of Blue Cheese on this trip. Can you speak Danish?"

"Well, er, no. But no need to, really. English is the second language of the Danes. Of course, fundamentally, the Danish language is quite simple. It's a mixture. All you need is some English, which you have, and some German, which I suppose you have, no? Of course, a little Anglo-Saxon does no harm."

"Heard about that stuff once in Eng. Lit. 101."

"Of course one should start with the Danish alphabet."

"Important to start right. Bullet: Start with the alphabet."

"Now do not confuse the O with the OE. Do not confuse the A with the AA. Or what would be worse, the A or the E with the diphthong AE."

"Will do. I mean, will not do. Bullet: Don't confuse."

"And there are very few Danish words, really, that begin with C. Most of them are foreign and easy to understand. And practically no words that begin with Q."

Pete Williams nodded assent. "I'll forget the Q. Makes life easier."

"And no W at all."

"Economical. Makes Danish a pushover, doesn't it?"

"Yes. Now in learning Danish, one could, of course, go the route of the primers."

"Maybe. I was planning to go the route of the Berlitz tape:

Er det et glas?
Ja, det er et glas.
Er det en hoene?
Ja, det er en hoene."

"I don't recommend that sort of thing at all, Mr. Williams."
"You don't?"
"No. You see those are all small words. The Danes have a strong tendency to pile up small words into one large word. You need to be bold and take a linguistic axe to the long word. Split the construction up into short words, and there you are. I'll show you how it works."

"Bullet: Be bold. Split construction."

Lucas Fysst reached into the seat pocket in front of him and pulled out a Danish publication that was tucked into it. He searched through the pages.

"Yes. Let's see. Here we are. This will do splendidly:

EJENDOMSMAEGLERFIRMAET

Looks formidable. The eyes can hardly grasp it all."

"Like the letters you get in a Scrabble game," Pete Williams confirmed.

"Now what hits me first—intuition, I suppose—is DOM. Now in Latin, DOMUS is 'house'. Yes, I suppose a bit of Latin helps occasionally . . ."

"Bullet: Latin helps."

" . . . so that leaves EJEN dangling in front."

"Dangle. Bullet: Danes dangle."

"Supposing that the J may be related to a G, one would have EGEN. Which is close enough to the German EIGEN for 'self' or 'one's own.' Thus we have 'one's own house.' One's own house is one's property. And that fits in with the FIRMAET at the end, supposing that word is allied to the English FIRM and German FIRMA.

"This leaves MAEGLER, and there is where German really comes in handy. MACKLER, in German, means 'middle man.' So, putting it all together, 'one's–own–house–middle–man–firm.' In short, 'estate agency,' or as you Americans put it, 'real estate agency.' You see how simple?"

"Absolutely lollipops. But funny thing: As you were figuring it out, I used the same method and came up with a different answer."

"Did you really? What did you get?"

"It's a firm warning not to smuggle meat products into the country. Makes sense. Denmark is a great meat producer. Keep out the competition. Europe not unified commercially yet."

"Remarkable! How did you ever arrive at that?"

"Easy. EIJEN: 'inside.' DOM: 'don't.' SMAEGLER: 'smuggler.' FIRM: 'firmly.' MAET: 'meat.' Roll them together: 'Inside–firmly –don't–smuggle–meat!'"

On deplaning, Pete Williams pulled out a card and handed it to Lucas Fysst. "Now remember, pal. If you're ever in Texas, just hop down to the Big Bend country, and Revelda Mae and yours truly will really show you a ding ding."

"Yes. Well, thanks, really. And should you come to Copenhagen, I'll be at Nyhavn 18. Next three months. Do visit."

Pete Williams pulled his notebook computer out of his pocket and punched in this information.

. . .

The discussions Lucas Fysst had at the University of Hamburg with Professor Dr. Dr. h.c. Topfenmeyer were totally irrelevant to the remarkable series of events detailed in this little book and will mercifully be omitted. These events changed humanity's understanding of the soul and also changed the posthumous reputation of one man. They are much too important to be muddied by discussions of Topfenmeyer's controversial interpretation of a first-century text.

. . .

On the train from Hamburg to Copenhagen, a trip of some hours, Lucas and Stephanie Eriksdottir had seats together. For a while, Stephanie slept. Lucas read and then nodded off. From time to time, they were awake together.

"Where will you live, Dr. Fysst?" Stephanie asked.

"Nyhavn 18. Near the harbour. You know, Hans Christian Andersen lived there years ago. In the very same apartment that I'll be staying in. Do you know Andersen's fairy tales?"

"I know his stories very well."

"And so do I. And you know, living there might infuse me with his spirit. I might write a story."

"Do you write stories?"

"Yes. When the spirit moves me. I write stories suggested by my cat, Thomas Gray. Well, she's not my cat, really. She's the College Cat. Very independent, she is. But I have a certain—what should I call it?—relationship with her. Smart as a whip, is Thomas. I simply play Watson to her Sherlock Holmes, Boswell to her Dr. Johnson."

Stephanie looked quizzical. Though her English was good, it did not seem to include a knowledge of Sherlock Holmes and Dr. John Watson, let alone the formidable Dr. Johnson and his biographer.

"You know, I don't think Hans Christian Andersen liked cats," Stephanie said.

"He didn't? Hmm. I've never really thought about the matter. You mean there are no cats in his stories?"

"Well, there are a few cats, but they're not important. And they're rather stupid. They say silly things."

"Really? I'm surprised. Can you give me an instance?"

"Well, there's a tom cat in "The Ugly Duckling." And he's not very important. He has a friend who is a chicken. What cat would have a chicken for a friend? And a cluck of a chicken, at that. They're a stupid pair. 'We two against the world,' says the chicken.

"And then the cat asks the duckling whether the duckling can make its fur spark, like his, or raise its fur, as he does. If not, says the cat, your opinions are fairly worthless. The duckling, who is much smarter, asks the cat in return, Can you swim like I can?"

"Yes, I see what you mean," Lucas responded. "That won't do now, will it? Thomas Gray wouldn't like that at all. What kind of a reputation would Andersen have if it got around that he was a catophobe? No, one wouldn't want that. The repercussions would be considerable. Might lead to an unpleasant revisionist assessment of Andersen."

The conversation lagged while the train stopped at a station and people got off and on.

"You know, Stephanie," Lucas said when the train was rolling once again, "I shall take the matter up with Thomas Gray when I get back to Pembroke College. She will be distressed. Indeed she will. Being a cat, she is naturally very pro-cat. Very appropriate. I'm sure she'll suggest something to rehabilitate Andersen's reputation. She is very good in the literary line—excellent in handling manuscripts. She had a great triumph a few years ago."

Stephanie, who did not quite know what to make of this, replied, "Why don't you write an Andersen story that has a smart cat in it? That would help his reputation."

"Are you suggesting a forgery? That would never do. Well, not an out-and-out forgery. A bit of ambiguity, perhaps? Creative ambiguity, perhaps? 'In the style of,' perhaps? Literary loopholes, perhaps? Where there is life there must be rules, and where there are rules there must be loopholes.

"Well, when I get back to Cambridge, I'll take it up with Thomas Gray. I suppose she's around somewhere, but I haven't seen her in a while. The Porter told me she's taken to living behind the potting shed in the Master's garden. Yes. Hmmm."

At four in the afternoon, the train pulled into Hovedbanegaard, the main railroad station in Copenhagen. As Lucas and Stephanie said goodbye, Stephanie said to him, "You know, Dr. Fysst, you look a bit like Andersen."

"Do I now? Hmm. Well, what do you think of that?"

Lucas went to the information window to find out how to get to Vallensbaek, a suburb where he was to be put up with a Danish colleague since the flat in Nyhavn 18 wouldn't be ready for a few days. Feeling quite adequately versed in written Danish, he intended to try out his spoken Danish. He asked his question, and the clerk answered it. Lucas jumped on a train and ten stops later ended up in Vedbaek, twenty miles from his goal.

"Lost in Copenhagen," thought Lucas. "And so soon. While I was watching the station stops, I must have forgotten my glottal stop. That's what did it."

The train back to the city was not due for twenty minutes. He walked to the nearby village and looked at the pictures of houses in the window of the local EJENDOMSMAEGLERFIRMA. Krone for pound, he said to himself, the prices of real estate look quite reasonable.

3

The Conference

Three hundred years ago, King Christian V, expanding the harbor of Copenhagen, built a cul-de-sac canal extending into the city perpendicularly from the existing waterfront. The canal is about four city blocks long and four or five fishing boats wide. It is spanned by one small drawbridge. Lined up in the Nyhavn (New Harbor) canal—it has been called that since Christian's day—are small fishing boats, excursion boats, a theater boat, a restaurant and cabaret boat, and Fyrskib Nr. XVII (Lightship No. 17), now retired from active sea duty and moored permanently as a kind of museum.

At the harbour end of the canal are several high-speed ferries to Malmoe, Sweden. A bit to the north, on Sankt Annae Plads, is the place where the large overnight ferries to Oslo dock. Within the memory of many Copenhageners, Nyhavn was where sailors, drunks, and women of questionable occupation were to be found. It has since become gentrified and touristified, with many fine restaurants attracting steady patronage. When the weather is warm, the restaurateurs set chairs, umbrellas, and bars outside for the swarms of visitors. The only remnants of Nyhavn's former identity are two tattoo shops and a seamen's bethel.

At the head of the canal is the largest square in Copenhagen, Kongens Nytorv (Royal Newmarket). In the center of the square is a statue of Christian V sitting on his horse, and surrounding Christian are hotels, department stores, the French Embassy, the Art Academy, the Royal Theater, and numerous sausage and flower stands.

On the south side of Nyhavn, halfway to the ferry boats and next to the Black Raven Restaurant, stands Nyhavn 18, a dark-yellow eighteenth-century structure of some six stories. The first story, one floor up from street level, was to be Lucas Fysst's place of residence during his stay in Copenhagen. This first story was where Hans Christian Andersen lived during the final years of his life.

After a few days of rest and an hour or so to pull his lecture notes together, Lucas was ready for his conference.

Det Kongelige Danske Filosofiske Selskab, the Royal Danish Philosophical Society, is a brisk twenty-minute walk from Nyhavn 18. You walk down the Gothersgade. You go past the gate of the Rosenborg Castle where a sentry is always on duty and soldiers are often drilling in the yard. You cross the Noerre Voldgade and walk along the Botanisk Have and, as you approach the lakes, there it is. Or, if you prefer, you can take a number 7 bus from Kongens Nytorv. It's a short ride.

The Conference

The dark-red building, though moderate in size, expresses pride in its heritage. It dates from the 1870s and was built in the heavy, florid architectural style of the times, which expressed solid confidence in the status quo. The building contains an office, a library, and a lecture hall. In the lecture hall, on elevated shelves between gothic windows sit busts of classical philosophers. On the left side, as one faces the stage, Heraclitus, Socrates, Plato and Aristotle look down on the audience with magisterial authority. On the right side, Thomas Aquinas, Pascal, Descartes, and Leibniz complete the philosophical ring.

Somewhat below the busts, as if in modest deference to the masters, there are framed pictures of the great men of Danish philosophy. Here can be found pictures of Sneedorff, Oersted, Steffers, Nielsen, Kierkegaard, and Hoeffding—men who, with the exception of Kierkegaard, are little known across the seas but who have contributed mightily to the firm bases of Danish thought.

The participants in the Conference on the Nature of the Soul, all wearing their name tags, filed in slowly and randomly and took seats. To the chagrin of the organizers and the editor of the conference proceedings, several of the speakers were very late in sending in their summaries. Besides this, there was a total misunderstanding and confusion as to whether the Swami intended to submit a statement or whether he even intended to appear.

A few words are recorded here about what went on in the presentations and discussions. They are derived from notes taken by Lucas Fysst.

Much was made of the soul as sign. One speaker went further: He converted the sign into numbers and the numbers into bits. In this way, a new concept of "soul bits" was introduced, crystallized, and elaborated.

The concept of the soul-in-itself was developed as an outgrowth of the more concrete notion of the cat-in-itself. Why, Lucas asked himself, are cats employed as thought experiments in theoretical physics and as illustrations in philosophical arguments? Why not mice or crickets? He recalled that this practice always annoyed Thomas Gray.

There was a rather heated argument as to the sense in which the relation "has" must be considered as "has-in-itself." (Philosophers are far from calm people. To philosophize is to argue, to refute, to demolish; ultimately to establish—at any rate, to establish for a year or two, until another view is born that slashes its way into the philosophic consciousness.)

There were talks on the soul as interpreted by the school of iterative naive empiricism and the soul as seen from the point of view of postmodern self-referentialism.

A morning and the first half of an afternoon were devoted to the soul viewed folkloristically. The name of Carl Jung was cited rather often.

A participant from Scotland brought up the treatment of animals in the stories of Hans Christian Andersen. "It is clear," he said, "that Andersen ascribed souls to all his animals."

That statement was disputed by a Danish participant. "It is my perception that such a sweeping statement, which refers to all animals in Hans Christian Andersen, must be modified to mean some animals in Andersen." The Dane cited two lines of Andersen's poetry by way of a proof-text.

> "Katten staar og krummer ryg
> men forstyrres af en myg.

> "The cat stands and makes a curved back
> but is disturbed by a mosquito.

"All a cat can do is to arch his back and paw at a mosquito? No emotion, no feeling? No reason? Only mechanical understanding? An automaton could do as much."

On the morning of the third day, the conference was picketed peacefully by a man bearing a sign in Danish. Translated, the sign read, "Each loaf of bread equals ten billion yeast plants killed." The picketer was passed off as someone operating on an irrelevant metalevel, and the conference went into scheduled presentations on the immortality of the soul as viewed by prominent nineteenth- and twentieth-century authors. One of the views put forth was that Hans Christian Andersen, a man who rejected many of the dogmas of the Danish Church, adhered vigorously to its doctrine of immortality.

On the fourth day of the conference, Lucas Fysst gave a comprehensive survey of Western opinion, as he had promised Ian Plumbley in the Pembroke Combination Room. He started with Aristotle.

"In Aristotle's great book on souls, the *Peri Tos Psychos*," he began, "argh, argh . . . Yes. I must remind my audience that the Greek word *psykhe* means 'soul.' Today's psychoanalysts are really people who untie the knots of the soul. Aristotle distinguishes three kinds of soul: the nutritive, the passionate, and the rational. They are located in the stomach, the heart, and the mind, respectively. Now animals most certainly possess nutritive and passionate souls. What is to be considered is whether and to what degree animals possess rational souls." The proceedings of the Copenhagen conference will soon become available, so there is, in point of fact, no need to record the remainder of Lucas Fysst's paper; this little book does not need to serve as an additional archive for the conference. Following, however, are extracts from the question period of Lucas Fysst's talk.

Q: Dr. Fysst, wouldn't you agree that your assertion of the uniqueness of the soul is incompatible with your assertion that the soul ultimately dwells in the plenum of all experiences?

A: I think the incompatibility is removed if one ceases to confuse the soul with experience. The soul dwells *in* the plenum but is not *of* the plenum.

Q: Dr. Fysst, what exactly are your views on Andersen's views on cats?

A: The evidence is moot. My intuition tells me that Andersen was more generous to cats than would appear, if not in his standard published work, then surely in his Nachlass, his residuary material. I shall consult Professor Elias Bredsdorff, my quondam fellow Cantabrigian at Peterhouse College, who is the author of the definitive biography of Andersen. I intend to look into the matter in some depth. I hope I shall be able to augment the published version of my talk with new information on this point.

Q: Would the speaker make clear and precise his own views on whether cats have souls?

A: This conference, you know, was inspired by the necessity of clarifying the larger question. Well, I thought I had made my position clear in my paper. If not, I shall reformulate it briefly:

The soul is a mental construct. It is not a physical thing such as a er, er, well, such as a cat. Mental constructs assuredly have existence. For example, the numbers one, two, three are mental constructs. The unicorn is a mental construct. Mental constructs can have well-defined features; some numbers are even, some are odd. We all agree to that. Or: A two-year-old can select a unicorn out of a random assemblage of pictures.

Mental constructs can attach to physical objects. Example: There are thirty-three participants at this conference.

The same goes for the soul. The soul is that which fuses what is perceived with what is conceived, or the particular with the universal. It can be attached to, or correlated with, a cat.

Now this is all metaphysics. There is also experience. Does my cat, Thomas Gray, have a soul? She does indeed, and five minutes with her should suffice to convince any unbeliever.

The corridor small talk over coffee and pastry was even more enlightening.

Dr. A: Do you know Guldenschlager's work? He speaks at 15:00.

Prof. B: Oh yes. You should read his *Untergang der Abendlandespizzakonzept*. That is, if you like political irony.

Dr. A: There is something intrinsically funny about the abstract pizza concept. There is something intrinsically funny about eggs. The physical egg, I mean. Has this been analyzed?

Prof B: Pizza has become an Ur-symbol.

Fr. Dr. C: Does the Swami talk? Has he arrived yet?

Msgr. D: The Svamin? No. I don't think he's arrived yet. He may not make the conference at all. Give a paper? Are you joking? That's not his way. He works one on one. One on three, maybe. But he's quite approachable, you know.

Dr. E: What do you make of Fysst's concept of *anima-in-posse*?

Dr. F: Mediaeval. Would you expect anything else of him? Besides, I think he ought to distinguish between *anima-in-esse* and *anima-in-actu*.

Dr. E: You know, Fysst works with this cat, Thomas Gray. When he deals with textual material.

Dr. F: His *anima-in-pussy*, I daresay.

G (a graduate student from Berkeley): On the basis of Dynelskaja's work on the concept of the enclave, would it be safe to assume the essential alienation of an enclave?

H (a graduate student from Aarhus): Within her context, of course.

G: Of course.

The Conference on the Nature of the Soul wound down. Goodbyes were said. Promises were made to send references and to exchange offprints of published papers. But during the conference Lucas's pro-cat sentiment had wound up.

4

The Search Begins

The search was now on for residuary material that would establish Hans Christian Andersen as pro-cat. Lucas Fysst scheduled a meeting with Dr. Sven Ahlborg, Curator of the Andersen Collection at Det Kongelige Bibliotek, the Royal Library.

The Royal Library is an easy twenty-minute walk from Nyhavn 18 and lies an area dominated by the Christiansborg Palace. The Folketing—the Danish parliament—is here, and if you go across the courtyard and through an archway of the Folketing buildings, you come to the library. Going up the stairs, you come to an old-fashioned shelf-lined reading room with many carrels. You would never

guess from this room that the Royal Library houses several million volumes—that it is, in fact, one of the great libraries of the world.

Lucas Fysst walked into the catalogue room to check whether the library had any of the books he himself had written. He found two listed. Not a bad place, he thought, giving the computer terminal a little pat, but it could be better. Fysst and Gray? Gray's brilliant work on the Spiral of Theodorus not here? This won't do. It really won't. Well, I shall send them a copy.

He found Dr. Ahlborg's office. Lucas introduced himself and shook hands. He presented his credentials. Dr. Ahlborg looked at him suspiciously. Although Lucas knows books, loves books, lives books, breathes books, writes books, buys books, and looks bookish, Dr. Ahlborg still viewed him with suspicion. There is as much hostility between librarians and the users of libraries as there is between dogs and cats. Librarians are afraid that in any number of ways users will diminish their collections of books. Users—particularly sophisticated users—are afraid that librarians unconsciously consider their holdings as private property and discourage access to them.

"I am glad to know you, Dr. Fist."

"Fysst."

"Yes. Fysst. How can I help you?"

"I should like to know whether you are holding any unpublished material of H. C. Andersen."

"No. Every scrap of Andersen material has been published. In Danish, of course. Other than his fairy tales, very few things have been translated into English, even though he is the Danish author best known outside Denmark."

"Do you have all his known manuscripts in your collection?"

"I believe we have many of the important ones. Some of the fairy tales are in the Andersen Museum in Odense. There may be a few unimportant things in private collections."

"May I take a look at one of his manuscripts? Any convenient one would do."

"Why certainly. You may examine it on the desk next to mine. But I have to tell you that the manuscripts in this collection may not be photocopied except by an act of the Folketing."

Dr. Ahlborg left the room and after several minutes returned with a handsome red box, tied up in shoelace fashion. He locked the door to his office and pocketed the key. He opened the box lovingly and tenderly withdrew a portion of the manuscript of Andersen's autobiography.

Lucas Fysst examined the pages slowly and carefully, noting the paper, the ink, the handwriting, the frequency of crossings out, the string binding, and the seals of the Kongelige Bibliotek.

"Thank you Dr. Ahlborg. Do you have a catalog of the Anderseniana?"

"Yes, but it is not complete."

"Perhaps you can help me. I have a suspicion—an intuition, if you will—that Andersen wrote a fairy tale that has remained unpublished. The manuscript has been lost. I should like to find it."

"It is not possible. Every scrap this man wrote has been published. What you suggest is very, very unlikely. If you found such a thing it would make a considerable stir, I tell you. I have no idea how to search. But I tell you more. Go to the dealer Viggo Pedersen og Soen, the Antikvariat, Fiolstraede. Ask about the private sector. Sometimes dealers know things that we librarians do not. And good luck to you, Dr. Fysst. Did you note the exhibition of Mozart manuscripts in the cases as you entered? Farvel."

"Thank you so very much. I will inquire at the dealers, as you suggest. Here is my address and phone number. If you think of something, please ring me up. Farvel."

The Fiolstraede, about a twenty-minute walk from the Kongelige Bibliotek, is lined with discount shoe stores, felafel shops, and second-hand book stores. The shops are full of signs reading TILBUD (sale) and OPHOERS (closeouts). The street is closed to general auto traffic, and the flow of pedestrians is substantial. Lucas found Pedersen's store near the Old University, where Krystalgade crosses Fiolstraede.

Viggo Pedersen, it turned out, had died in 1904. His son, Viggo Segers Pedersen, had died in 1947. The shop was now owned by a certain Leif Anders Knudsen but run by Knudsen's aunt, Froeken Nielsen, a lady in her sixties.

"Do you handle manuscripts, Froeken Nielsen?"

"We have some. Are you a collector?"

"Well, yes and no. I have something specific in mind. I am looking for Anderseniana."

"Very, very rare, by now. All in private hands or in the Kongelige Bibliotek or in the Andersen Museum in Odense. The University Library in Aarhus has a few."

"Surely private libraries are sold from time to time?"

"We handle estates, but I haven't seen anything by Andersen in one of them for years. Well, the odd letter. Nothing major. Now, would you be in the market for some Blixeniana?"

"Thank you, but no thank you."

"We have quite a bit. I suspect that Blixeniana is being produced in great quantities."

"It shouldn't surprise me. Where there's a market there's always a supplier. Can you suggest another dealer who might know something?"

"This part of Copenhagen is full of old book dealers. Just wander around. Go in and ask."

"Well, here's my address and phone number. If something should materialize, please ring me up."

Apart from random inquiry, Lucas had one more lead to check out that day. On his way, he passed a coffee and pastry shop. He pressed his nose against the window. The lead could wait.

5

Conversation with a Bouillabaisse

The search for a lost Andersen manuscript—a presumptive lost manuscript—was not going well. Random inquiry with book dealers led to raised eyebrows and a polite cup of coffee but nothing more. One dealer suggested that Lucas contact a certain Helger Helgerssen, an heir to beer and shipping fortunes, who was a collector of first editions. The dealer said that Helgerssen was a very knowledgeable and accommodating man.

Lucas rang up the number provided and learned through a secretary that Helgerssen was in the Odsherred peninsula, where he had a villa on the sea.

"When you speak to him next," Lucas said to the secretary, "I'd be greatly obliged if you would ask him about the possibility of unpublished Andersen manuscripts. Let me give you my number in Copenhagen. The name is Fysst. Yes. F-y-s-s-t."

. . .

After a day spent in futile search, Lucas determined to put the matter out of his mind for a bit. He walked back to Nyhavn 18. The weather was murky, wet, windy, and cool. It was already dark. He stopped at the IRMA supermarket on the harbourside Havnegade to pick up a few items for supper.

Arriving back exhausted, he fell asleep. When he woke up, the hands of the old brass ship's clock in his bedroom were moving toward 7:30 P.M. Lucas Fysst's stomach rumbled. He began to soliloquize.

"It's now 7:30 P.M.," Lucas said to empty space, "or, in local terms, 19:30. To add A.M. would be wrong and to add P.M. would be superfluous. But 19:30 does not match the clock, which points to 7:30. If the Europeans were as logical as they claim to be, why don't they make a clock whose hour hand rotates once a day? Ah, I suppose it's these little inconsistencies that make life possible."

Lucas went to the refrigerator and opened it. "A carrot. Yes, I see a carrot. I see a carrot, you hear? The carrot will be the rock on which I shall build my supper. I see an onion. Splendid. An onion, what? Onion: Are you a genetically engineered onion, or an onion of the old school?

And an answer came there from the onion: "Shall the tears be removed from life?"

"What more can I find to add substance to the nutritive soul and sentiment to the passionate soul? I see half a green pepper. Yes, a green pepper would be consistent with an onion and a carrot. A potato perhaps? Don't hide from me, potato, I know you're in there somewhere. Come, let me palpate thy convexity, potato. Thy too too solid flesh shall melt and resolve itself into a stew.

"Yes, old potato, a conception is growing; a dish is taking shape in my mind and will soon be instantiated in the pot. A noble dish will arise where previously there was nothing.

"*The Paratrimsika* put it this way: The great banyan tree is contained in the seed; *in posse*, of course. In this way the entire universe, with all its movable and unmovable beings, lies potential in the heart of the supreme being.

"What else does it need? A pinch of this and a crumble of that. A bit of water. To cover. As it says in the book of *Isaiah*, 'as the waters cover the sea.' Set it on to simmer.

"Hmm, I sense incompleteness. Aha. It's just as well I bought a small piece of torsk. It's in the shopping bag. 'What is torsk?' You may well ask. Torsk, to us Cantabrigians, is cod. You agree? But who or what is this? Salmon? I shall now tip in this tin of salmon. Laks to the locals, caviar to the general. Thereafter to simmer slowly. Like a Dane when enraged. Such is the Danish character, I've been informed."

Forty minutes later, at the table, a dialogue ensued.

> LUCAS FYSST: Soup, may I taste you?
>
> SOUP: Please. Be my guest.
>
> L.F.: Hmm. Splendid. You are an alimentary joy. You are a bouillabaisse.
>
> SOUP: If you say so, sir. However, I do not feel myself to be a bouillabaisse. I feel myself rather as a potage.

L.F: You are much too aristocratic for a potage, O noble bouillabaisse.

Soup: Tell me. If I'm a bouillabaisse, what is my genealogy? Am I related to the Bouillabaisses of the Ancien Régime? How was I born?

L.F: How were you born? Don't embarrass me. I'm quite Victorian, you know.

Soup: Come, come, this is the twenty-first century, nearly. You can't shock me.

L.F.: Well. How shall I put it? You're just a little something I tossed off.

Soup: You mean I'm not the first? There've been others? You go around tossing off soups, do you?

L.F: I admit it. There've been others. A mulligatawny in Cambridge. A bisque in Boston. The urge arises once in a while.

Soup: Something lacks, sir. Though flattered, I feel myself manqué for a bouillabaisse. A wee bit of garlic? A small piece of crab or langouste? The odd clam?

L.F: Do not speak to me of langouste. We live in the Land of Sild and Torsk. I deem thee a bouillabaisse, and there's an end to it. Mastery reveals itself in limitation. So wrote the great Goethe. I tell you that langouste is neither necessary nor sufficient for you to be deemed a bouillabaisse.

Soup: The odd clam?

L.F: Unnecessary. Do you know the story of Sir Roger Manning, O soup? No? Well, I shall tell you. Sir Roger was a distinguished oripologist, or something like that. At Oxford, you know. And Queens at Cambridge (or maybe it was Jesus College) wanted him badly. And he accepted their offer, but when he got there they wouldn't allow in his pet dog. You see, Queens (or perhaps Jesus College) had a rule prohibiting dogs.

Sir Roger said firmly, "No dog, no Manning." Wouldn't budge. Stubborn as a mule. So what to do? The Fellows of Queens (could it have been Jesus?) met in an extraordinary session. The Master, splendid chap, really splendid, head screwed on right, spoke up.

"Since 1653, when the Senior Preceptor was bitten and died of distemper, this College has had a rule to exclude dogs. However, this College has never excluded cats. Never. I shall now put it to the Fellows of this College that Sir Roger Manning's dog be deemed a cat. Those in favour will respond by saying 'placet.' Those against will say 'non placet.'"

And placet, placet, placet it was, all the way. What would Thomas Gray have thought of that: a dog deemed to be a cat?

Soup: I feel increasingly limited. In quantity.

L.F. (spooning considerable quantities into his mouth): Though surely not in bouillabaissekeit.

Soup: I go now. I am more and more limited. I vanish. I evanesce. I fade. My raison d'être is fulfilled . . .

L.F.: You see? What did I tell you? A mulligatawny would never speak French. Hmmm. What am I hearing? An evanescent soup speaking French? What am I saying? Am I hallucinating? What's overcoming me? Is life too hard; the tasks thereof too complex? Oh Thomas, I wish thou wert here beside me in Copenhagen. We should show the world a trick or two then. You and I and the world, Thomas.

6

Thomas Gray Arrives on the Scene

There is a general opinion that cats dislike water. Regarding bathing and romping in the surf, this is most certainly the case. Regarding proximity to water, sailing, and standing on deck observing the droplets of spray and questioning whether each drop has an individuality, this is most certainly not the case. Thomas Gray is living proof.

At the very moment when Lucas Fysst was conversing with a soup of dubious pedigree, Thomas was on the upper deck of the *Jewel of Anglia*, a ferryboat that plied between Harwich and Oslo. The air was cold, the wind and sea were calm, and the stars were bright. Every now and then, a wavelet far beyond the wake reflected one of the many lights on the deck.

These were moments Thomas loved—moments that displayed the empty surface of unwritten potentiality. The intimation of eternity was before her, and the solid past was behind. The prow cut the waters. The waters split and raced to the rear. They churned in agitated spirals. The turbulence gradually abated and disappeared. The memory of the passing ship was damped out, erased.

Why is there something and not nothing? Empty space and empty time: Can one imagine such things? Does not space cry out for occupation? Does not time stand bereft without event? The union of time and space is a manifestation of cosmic desire. Creation begins in violation, violation of the purity of nothingness.

Suddenly, Thomas Gray terminated her reverie. Her whiskers twitched violently and her paws trembled. Though she felt pulled forward, she questioned her own goallessness. Lucas is in deep trouble, she said to herself. I know it. I feel it in my paws. He is looking for something and he can't find it. He is in over his head. He needs me. I shall go to him. When I reach Oslo, I shall book immediate passage to Copenhagen and look for him. When apart, we are independent. Now that's a noble state but an incomplete state. When together, we are the world.

. . .

The *King of Scandinavia*, a huge white ferryboat that plies regularly between Oslo and Copenhagen, pulled into its slip in Copenhagen and discharged its passengers, autos, and huge lorries laden with goods. At the end of the procession, a medium-sized cat walked off and took her first look at the new country. Sure of her goal but unsure of her location (well in accord with the indeterminacy principle of quantum physics, whose philosophy was thrashed out in Copenhagen in the 1920s), Thomas Gray raised her tail high and set off to find Lucas Fysst. There ensued two days of wandering—days that were entertaining in themselves, with, as she allowed much later, no cause whatever for panic.

It is hard now to document the exact circuit taken by Thomas Gray during those two days, but from tangential remarks made later, it seems to have been as follows: From the Admiral Hotel, where the *King of Scandinavia* docked, she wandered along the harbourside. Being then in a state of semidisorientation, she found herself on the fast boat to Malmoe, Sweden. This error corrected, she went back to the Admiral. Thence, through passages not available to the general public, she made her way west to the courtyard of the Great Marble Church. The church door was locked. Had it been open, she might have relaxed her traditional stance of avoiding monumental buildings and gone in. Instead, she spent some time looking at the statues of the Fathers of the Danish Church outside the building. It is a measure of the general tolerance of the Danes that Soeren Kierkegaard, who in his day was surely an anti-Father, is to be found among them.

Wandering still farther west, toward the Great Royal Road, she found a sausage stand and a discarded sausage-en-roll on a public bench. She dispatched this without critical comment. Then, feeling sufficiently if not marvelously fed, she went back to the bottom of Kierkegaard's pedestal, curled up, and fell asleep.

For lack of information, she might have wandered indefinitely, lost in Copenhagen, unable by random roving to locate Lucas Fysst.

But after she woke up in the grey-yellow, windy, foggy morning, she happened upon some information. Walking east, Thomas came to the central square of the Amalienborg Palace. In the distance, near the ceremonial chains that circumscribe an equestrian statue, she noticed a grey-yellow misty cat. A conversation ensued.

"Greetings and goddag. I'm Thomas Gray from Cambridge and Waterfen St. Willow, England, you know. I seem to be lost in Copenhagen. You from around here?"

"Goddag. I'm Mewsli. Yes, I'm from around here, so to speak. I don't normally indulge in witticisms when the sun has hardly risen, but these are my purlieus. Are you are a tourist?"

"No. Heaven defend us all from that category. I'm here on a consulting job of an academic type."

"Well, I try to stay away from the world of ideas, so spare me the details. I'm in the social line myself. You're looking for a place to stay?"

"Right. Perhaps more pertinently, I'm looking for my collaborator, Lucas Fysst. English chap. He's not been in Copenhagen too long. Have you heard of him ?"

"The name sounds vaguely familiar. Yours also. But I tell you: Copenhagen is a city with several million souls. We can't keep track of all the comings and goings, can we? What sort of accommodations is he likely to have booked?"

"Well, back in Waterfen St. Willow, we live not too far from a canal. I should think he would go for a view of the water."

"There's a canal very close by. Nyhavn. Start there."

"Will do. One more request, and then I'm off. It's been several days and I don't have my land legs yet. I'm still a bit wobbly. Could you recommend a pied à mer for a couple of weeks? I enjoy gentle rocking wave motion."

"Nothing simpler. I suggest Fyrskib Nr. XVII on the Nyhavn. The concierge is one of us, a cat named Knurhaar. From Iceland. Tell her I sent you. Just go that way to Tolbodgade. At the Sailors' Bethel, turn right. Then over the canal on a little drawbridge. That puts you on Nyhavn. Turn right again. You can't miss the ship. You'll see a light on a tall mast. The Black Raven is adjacent. It's a restaurant of some integrity—at least I've been told as much. You can get lunch and supper there, but no breakfast. Good luck."

"Ta ta."

"Farvel!"

There is nothing better than asking to disambiguate matters, Thomas Gray thought. The world is divided into three types: those who always ask, those who never ask (these are principally males), and those who ask only under duress. And the world is further cross-divided into three additional types: those who believe what they are told, those who do not, and those whose belief or disbelief is haphazard. This makes nine possible policies and there are advantages to each. Asking one's way, thought Thomas Gray, is by no means an open-and-shut proposition. But I am glad I did it.

The day had become lighter and somewhat windier by the time Thomas walked across the drawbridge. At the Sailors' Bethel, she turned right. Past Nyhavn 18, she found the Black Raven, and moored opposite the restaurant was Fyrskib Nr. XVII. Thomas found Knurhaar, the concierge. She checked in and, without further conversation, promptly fell asleep inside a coil of mooring tackle.

7

The Collaboration Renewed

At this point in our story, having introduced in a substantial manner Lucas Fysst, one of the two principal characters, it is necessary to identify the second principal character more closely. This is for the benefit of readers who have not yet experienced the pleasures of *Thomas Gray: Philosopher Cat*.

Thomas Gray is a female cat—yes, female; readers who are interested in knowing why this is the case are directed to the book just mentioned, where all is made crystal clear. She is around five or six years old, smart as a whip, independent, often touchy, and sometimes grouchy. When at home in England, she is generally to be found at the bottom of the Bursar's staircase at Lucas Fysst's Pembroke College, where she occupies the Mat of College Cat. She is frequently on leave, sometimes at Lucas Fysst's house in Waterfen St. Willow, in the fen country of Cambridgeshire, not too far from where she was born. Equally often she is in the harbour areas north of Cambridge. She has a penchant for boats and travel by water, as she is descended from Dutch cats who crossed the water several centuries ago with Dutch families engaged to drain the fens.

A year before the present story opened, Thomas Gray earned a considerable reputation in learned circles by making sense out of an ancient mathematical manuscript. The *London Sunday Observer* picked up the story and sent a reporter to Pembroke College. Thomas was photographed cradled in the arms of the Senior Tutor of Pembroke, a position to which she objected vigorously, saying (and rightly so) that though the Senior Tutor was a worthy gentleman, he had nothing to do with the case. The picture came out in a two-page colour spread in the *Observer's* magazine section.

Achieving great fame and publicity at an early age is always a mixed blessing. Life can be downhill from there on, and immediately after the newspaper publicity, Thomas fell into ennui and depression. Luckily, she recovered, albeit somewhat slowly. Upon recovery, she pursued her work competently, if not spectacularly. Whenever she worked on scientific or literary matters (her specialty was manuscripts) it was in collaboration with Lucas Fysst. When she worked on philosophical matters, she was completely on her own.

The morning after Thomas Gray checked into Fyrskib Nr. XVII, Lucas Fysst turned the wrong way up Nyhavn. He had intended to go to the IRMA supermarket; instead, his error took him past the theater boat and toward the fireboat. Suddenly, he noticed a familiar figure sitting above the taffrail of the ship.

"Thomas, is that really you? What are you doing in Copenhagen, old girl?"

He ran toward the gangplank. Thomas Gray recognized her friend and, equally surprised, yet not surprised, she ran to meet him.

"Hello, Professor," she said, "Yes, *c'est moi*. But here, they call me Tomas Graa." She puckered up her nose as if to pronounce the strange vowels. "And I ask the same of you. What are you doing in Copenhagen? It's been months since we've been together. Well, think of that. We cats have a saying: 'One must wander abroad to meet one's neighbors.'"

It will be noted here that cats speak neither English nor French nor Danish. When with one another, they speak in a mutually comprehensible language that is not always vocal. When Thomas and Lucas are together, Lucas quite often speaks English, though Latin and Greek also come out a bit, and Thomas answers in a mutual language of looks, actions, and sounds that, after years of tender association, are perfectly intelligible to Lucas. The words recorded in this volume are his transcriptions—his English subtitles, if you will.

"Well, you see, I came for an international conference on the nature of the soul. I gave a survey paper, Western opinion largely, on whether animals have souls. What's your opinion on the matter? Do animals have souls? I'm staying on a bit. Several months. Depends on the funds. I'm finding Denmark a bit pricey. And you, Thomas?"

"Well, now that you're here, I think I shall also stay on. No problem whatever with the funds. Where are you staying?"

"Nyhavn 18. That's my address. Just two houses up. They've given me a sumptuous flat on the first floor. You've heard of Hans Christian Andersen, have you? He lived in my flat. Years ago. You must visit me from time to time."

The game wherein Lucas Fysst and Thomas Gray pretended to be surprised by each other's presence in Copenhagen was totally understandable. After all, each had feelings of independence to maintain. Once the little game was over, once the sense of coincidence had transformed itself into the sense of intuitive and purposeful action, they got down to business.

Lucas told Thomas how a rumor was spreading, a rumor that had to be stopped, that Hans Christian Andersen had been anti-cat. Lucas told Thomas that he suspected Andersen had left behind an unknown, unpublished manuscript in which a cat was viewed quite favorably. Lucas said that he was on the trail of such a manuscript and so far had had no luck. He enlisted Thomas's help in the search.

Thomas fell in with this project with considerable enthusiasm, and the reasons for this enthusiasm stemmed from her recent fame. Shortly after Thomas's great success with the ibn Kurra manuscript, and after her considerable publicity in the *Sunday Observer*, the Senior Tutor at Pembroke College received a quantity of mail, the subject of which was Thomas Gray. Would Thomas Gray consent to go on a Brains Trust TV show? Would Thomas Gray endorse Kittikid, the feline equivalent of steak and kidney pie? The pay would be a lifetime supply of the stuff. Could Thomas Gray, by her intuitive or clairvoyant powers, find a wedding ring that had been lost and whose loss was now causing a domestic rift in Somerset?

Thomas Gray heard about all these requests and was flattered by their existence, but she was irritated that they had been addressed to the Senior Tutor and not to herself. That, she thought, would have been the decent thing to do, as the Senior Tutor had played no role whatever in her professional life. At the very least, they should have been addressed to Dr. Lucas Fysst, who was her coworker.

And then one day, in the midst of the constant stream of letters, an unusually nasty letter came to Lucas, which read as follows:

> *Your cat, Thomas Gray, is a complete fraud. Cats cannot think properly. I know from bitter experience. My gray and black cat, Lulu, knocked a soft-boiled egg off my breakfast table and spoiled the rug. Was that thinking?*
>
> *Yours faithfully,*
> *(Name withheld)*

Digesting the letter slowly, Thomas got hotter and hotter under her fur. Cats can't think properly? What a lie! What a deliberate canard! What sensible person would ever place a rug under a breakfast table? Was that thinking? The mess from an egg? Look at the mess the World Egg is in. Look at the human mess. Is that an instance of rational thinking? Of intuitive thinking? Of romantic thinking? Is that what is called human thinking?

False ideas need to be vented, discussed, argued, and then extirpated. Thomas was ready to counterattack. The rehabilitation of Andersen's reputation with regard to cats was a minor step, but it was a step forward, and she would do what she could to bring it about.

"I suggest we work independently for a while," said Lucas Fysst. "I've already started with libraries and old book dealers and auctioneers. There are numbers of them on the Fiolstraede not far from here. I shall turn next to private collectors. You have your own methods, of course."

"Of course."

"Good. Then let's get on with it, Thomas."

"Let's. Well, ta ta, Professor. I'll come around to Nyhavn 18 once I've settled in."

"Toodle-oo."

Lucas turned and toddled off, this time in the correct direction for the IRMA supermarket, and made a mental note to lay in a supply of Whiskas. Thomas went back to her place on the taffrail of Fyrskib Nr. XVII, thinking that although the lightship offered all conceivable modern conveniences, catwise, an occasional snooze at Nyhavn 18, first story, might break the monotony of gentle wave motion. Then, turning from the trivial to the more substantial, she began to plot her course of action in the collaboration.

8

Found on Stroeget

Stroeget is a long, narrow shopping street in Old Copenhagen. Actually, it comprises five separate streets joined together into one continuous path that winds its way from Raadhuspladsen (City Hall Square) to Kongens Nytorv. Its shops run the gamut from the upmarket Georg Jensen's to Macdonald's, from bookstores and bars to ladies' fashions and fine chocolate, from florists and furniture to office equipment. Except for the occasional police car or delivery truck, Stroeget, like nearby Fiolstraede and Kobmagergade, is free of automobile traffic. The thick and turbulent current of pedestrians is sizeable. The presence of shoppers, tourists, drinkers, buskers, pantomimists, hurdy-gurdy men, and beggars make it one of the most popular and picturesque places in Copenhagen. The Old City itself is a maze of streets and little squares. Stroeget bisects it, and if the newcomer to Copenhagen were to wander just a few feet off Stroeget to take a quick look at the toy museum, for example, he might very well find himself lost.

That is exactly what happened to Lucas Fysst. He was on Niels Hemmingsens Gade, but he didn't know it. He consulted his *Kraks Bykort*, a thick atlas of Copenhagen and environs. He seemed to be located just where the binding joins the left- and right-hand pages of Krak, which made reading impossible. The wind ruffled the atlas pages. His reading glasses fogged over. Raindrops dampened the paper at location F-4. He looked up to where there should have been a street sign, but there wasn't one.

While Lucas was in this perilous condition, a short, dark man wearing a Nehru jacket and a loincloth approached him. The man carried a begging bowl to which had been taped a sign that read, "No coins, please." The man spoke.

"Dr. Fysst? Yes? I have been aspiring many years to make your acquaintance, and now you see destiny working in the very finest way."

Lucas was confused initially. Then, it all came together. "You are . . . Swami?"

"The same."

"I am delighted, so delighted," beamed Lucas. "You must be starving! Come to my flat. I shall give you nourishment. Where are you staying?"

"I stay in the Christiania."

"Is that a hotel nearby?"

"No. It is a neighborhood that expresses a new Tendency of Life. I have made new disciples in the Christiania, and they have given me nourishment."

"Are you limited to eating fallen fruit? You must not eat plucked fruit?"

"There are many impediments to travel. However, I have made application to Ganesha, who is our god in charge of dissolving impediments, and the ways and means have become clear."

"Come to my flat. I must hear your opinions on whether animals have souls. Let me take you there. But which way? Which way? Where is the main street?"

"The main street is just over there," Swami pointed. "But where is the main street one should take in this illusory world? The main street is to be found in the understanding of the *Siva Sutras*. Now I must hear your opinions on whether these sutras have influenced Kant."

After getting back onto Stroeget, Lucas led the Swami to Kongens Nytorv. They walked across the square.

"Mind the infernal cycles, Swami, they're often worse than the cars!"

"Ah yes. The Eternal Cycles. There is no way to avoid them. They are life itself."

Lucas sighed, and disjointedly added to the Swami's remarks an expression of his own perplexity. "Where is 'here'? And when is 'now'? Can the physicists answer us with assurance?"

They walked past the sausage stand and the statue of Christian V. My very goodness, thought Lucas to himself, how the kings of the world have become younger! From Christian V to Nyhavn 18 was only a few minutes.

"Ah, Swami," said Lucas Fysst, opening the door of his flat at Nyhavn 18, "I'm glad you were able to come. We must talk about the doctrine of ahimsa; indeed we must. We must talk about many things. Indeed."

Lucas led his visitor into a study that had on its door a small porcelain sign reading "H. C. Andersen." The Swami took off his white jacket, revealing the Sacred Thread over his shoulder. He opened his travel bag and took out a tiger skin. He placed it carefully in the middle of the floor and sat down on it. He set his begging bowl to the side.

"The doctrine of ahimsa is the doctrine of nonviolence and the sacredness of all things that live," asserted the Swami in a loud voice. "If only its truth would spread in the world. How much tragedy could be avoided. Ahimsa is what the spirit becomes through self-realization.

"I have heard that your Aristotle, who knew something about souls—but not too much, I should assert—permitted the eating of meat. That was most unfortunate. I have just completed a poem on the matter in twelve thousand two hundred forty lines of Sanskrit."

Thomas Gray wandered into the room dreamily. She noticed the tiger skin and was electrified. She growled and then regained her composure.

"Swami, this is my cat, Thomas Gray. Well, not my cat, really, but my occasional collaborator."

"The cat is a greatly symbolic animal. I have the strong impression that the Buddha—I do not concentrate too much on the life of the Buddha—was never reincarnated as a cat. A rabbit, yes. But not a cat."

Thomas looked pleased at this remark. She wandered over to the begging bowl and examined it.

"But then, again, of course, all animals have their symbolic aspect," the Swami added.

Thomas heard this second pronouncement and considered it an equalization of all animals. This was a doctrine she did not adhere to. To her, cats were the Chosen Ones. She emitted a long wail of disapproval. The Swami looked at her. The cat examined the begging bowl further.

"I will tell you a story that will make you understand the equality of animals. The great king of the gods, Indra, was once proud and haughty. It was said that he was building a palace whose magnificence could hardly be imagined. One day he met Shiva, in the form of a little Brahmin boy. The boy rebuked him. The boy asked Indra to turn around. Indra did so and saw a long row of ants, hundreds of them, thousands of them, all plodding in single file, transporting material to their hill. The little boy said to Indra, 'Consider that row of ants. Each one of those ants was formerly an Indra.'"

Still loitering near the begging bowl, and hearing this story, Thomas Gray emitted an untranscribable noise.

"Stop that racket immediately, Thomas. We are trying to conduct a serious conversation."

"Oh no," said the Swami, "allow your cat to have its say. My word. I think . . . yes, I think . . ."

"As a rule, Swami, she's steady and quite sensible. Once in a while she goes off on a toot. She had an episode several months ago. She imagined she was sitting on one of the lower branches of the Yggdrasil. Do you know the Yggdrasil? No? Then I shall tell you.

The Yggdrasil is the World Tree, whose roots extend into the three universes of the gods, the giants, and humans.

"As I said, Thomas imagined she was sitting on a lower branch. She saw a squirrel running up and down, stirring up trouble between a dragon who was gnawing at the roots and an eagle who was on lookout at the top. She told me that she generally stood above the fray but that from time to time she felt she had to volunteer a bit of advice to the squirrel. Thomas is rather Northern in her mythic outlook, I should say."

The Swami's mind was elsewhere. Having hardly heard the story, he said, "O Great Cat, you have recognized my bowl, have you? That bowl was passed down to me by my revered teacher Vyaghrapada. Professor Fysst, this cat has recognized my bowl as its own bowl. This cat, I believe—I'm sure now—possesses the soul of my venerated master, Vyaghrapada. O my revered Guru, how wonderful is the world in its myriad manifestations. And you have taught me not to be in bondage to it."

Thomas Gray was confused. On the one hand she felt that her own unique identity was being stolen from her—that it was being transmuted into an alternate form. This was the negative side of the matter. On the other hand, her immediate elevation to guru status acted on her vanity as the smell of coffee in the morning acts on a sleepy human.

Proudly, and with tail high, she strutted out of the room as Lucas began to speak.

"How did you come to Copenhagen, Swami? I realize that for a man in your framework, there are impediments to travel. I believe you are not permitted to cross the waters. Copenhagen is on an island. How did you manage this?"

"Without impediments, life would be meaningless, and the cycle of many afterlives would be doubly or triply meaningless. That is a quarrel I have with your Western theology and your Western social meliorism. It aches for the Great Equilibrium of Perfection. The second coming. The messianic period. An equilibrium is a state without impediments and therefore a state in which the Eternal Cycle is terminated and denied. Human problems are the very breath of life.

"But you are right. It is not allowed that I cross the water. Before I set out, I petitioned the great god Ganesha who is the Dissolver of Impediments, and I received enlightenment. I traveled overland from Benares through what was once Soviet Land—everything dissolves—through Poland and Germany. I came then to this land of water and fjords. Where there is a bridge, its two sides are to be considered as forming one connected land. But you are perfectly correct, the isle of Sjaelland, where Copenhagen is, remains unconnected."

Thomas Gray reentered the room and resumed her place near the Swami's begging bowl.

"I was told that plans have been drawn to connect Sjaelland with the nearby island by a bridge. In my enlightenment, I saw that the distinction between what has been realized and what is destined to be realized is as nothing from the point of view of the endless cycles of infinite time. In this way, the impediment was dissolved."

"Splendid, Swami, splendid," said Lucas Fysst, rubbing his hands together. "We must deem; we must deem when the necessity arises for deeming. A dog is deemed a cat, an island is deemed mainland if connected by a mental bridge-to-be. It is the only humanly possible way. Language is too restrictive for people to be bound by its implications."

"But it is not deeming, my dear Sir Reverend Fysst. It is recognizing the essential unity of what appears to be disunited."

The word "deem," Thomas Gray agreed with Fysst, is the only antidote that language and law have against their own self-created rigidities.

"And now, Swami, down to a bit of business. I should like to engage you for private instruction in, say, the *Siva Sutras*."

"I am delighted that you have become a seeker of the truth, Reverend Fysst."

"I seek the connections between the religious impulse and the mathematics of the second-century *Sulvasutras*."

"I shall be honored. It is written in the Commentaries that whatever is perceived in an unilluminated way despite the attentiveness of Mind becomes clearer when the Will is exercised. From this it follows that whatever existed in whatever form or place or time becomes clear if the power of spanda is invoked. I shall come to your flat again, as we shall arrange. May there be benefit for us all."

9

The Glory of the Realm

Lucas Fysst, lecturing to Thomas Gray, who was sitting on two pillows, said, "Creative self-indulgence, that's the secret of life."

Agreed, Thomas Gray thought, assuming it's available, which it not always is.

Lucas went into the dining room and unwrapped a package that had printed on it:

> WIENERBAGERIET OG KONDITORIET
> KOEBMAGERGADE 27–29.

He extracted six small chocolate marzipan balls. He set them in a circle on a little plate.

"Two now. Two before supper. Two for dessert. Just what the doctor ordered."

Every day, as he came home along the Koebmagergade, Lucas passed a certain bakeshop. In the window, cakes and tarts and tortes and slices and crescents and whorls and whirls and balls and squares and rolls and strudels and petits fours and petits sixes were on display. And, to this window, Lucas Fysst was attracted in the way a bit of tissue is attracted to a comb. He would press his nose to the pane, stare indefinitely, and soliloquize.

"I have no desire to comment adversely on the pastries in my own Cambridge—in Fitzbillie's shop, to be specific. I think Fitz has done splendidly by me, having brought me to this point in life. But here we are in a different dimension of pastry altogether. The bakeshops, the pastry shops, the konditoreien of Denmark are surely the glory of the realm.

"Ah, the pastry. The pastry. I melt. My eyes dissolve in tears. I push my nose against the windows and stand transfixed in longing. The snegler, the spandauers, the aebletaerter, the tvindinger, the boller, the wienerbroed, the kager, the marcipan, the snitter, the horns, the Jenny Lind skiver, the Sara Bernhardt kugler, the jordbaer taerter, the kronprinsesseplumkager . . .

"Yes, the pastry. How wonderful it would be for a person to be memorialized in pastry! The Lucas Fysst Lectures at Cambridge University? Perhaps. The Lucas Fysst Chair of the History of Science? Possibly. But a hasselnoeddefloedelucassnitte! That would be living; indeed it would.

"Hasselnoeddefloedelucassnitte." At this point, Thomas Gray, who was in Lucas's shopping bag, sneezed. "Mark the word carefully. Repeat it over and over again. Use it as a mantra. Combine it alternately with the sacred syllable Om. The word cannot really be translated, neither into English nor Romanian nor Sanskrit. Roughly, it means hazelnut cream Lucas slice. A better translation: gateway to that portion of paradise where the material and the spiritual are united.

"Ah, the pastry. I want to do the mathematics of pastry, the rhetoric of pastry, the philosophy of pastry, the poetry of pastry. What is the relationship between the gods on Olympus and pastry? What? Is Olympian life all ambrosia and no pastry? The gods in Valhalla, in contrast, have surely seen matters otherwise and have laid on a paradise of pastry.

"But I must be practical. I must avoid the nemesis that necessarily attends the hubris of pastry. Self-indulgence, creative or otherwise, must go hand in hand with self-restraint, with frugality. In this way, the balance in the universe is not upset.

"I propose to adopt the following regimen. Each day I shall enter the Wienerkonditori and work my way across the display trays. From left to right, upper tier first, left to right, lower tier next, a different tray each day, I shall select and buy an appropriate number of whatever pastry it is. In this way, with balance and equipoise, without prejudice to any particular item, I shall have run through the gamut of the pastrycook's art several times before I leave Copenhagen. Two pastries, then, before supper; two with supper; two after supper, the last two to be accompanied by the thimblest little nip of Cherry Kijafa."

After supper that night, Lucas sat in a soft chair and began to study Book II of the *Etymologies* of Isidor of Seville, an early seventh-century scholar.

"Knowledge of numbers," he read in Isidor, "should not be held in contempt."

"No indeed," said Lucas a bit drowsily. "For example: two before supper. This should not be a contemptuous matter."

He read on. "In many passages of the Holy Scripture, the mysteries contained therein are explained."

Lucas agreed. "Verily. I take Isidor to mean here the mysteries of number."

More from Isidor of Seville: "Not in vain was it said in praise of God: You made everything in measure, in number, and in weight."

"Two after supper . . . ," Lucas responded to Isidor, and with that his eyes closed and he fell into a heavy sleep.

10

Spectral Dialogue in Nyhavn

It happened the next night at a late hour. Lucas, quite tired out from considerable travel on the buses of Copenhagen, sat at the desk in the Andersen room of his flat and looked out the window over the canal. Thomas Gray was on a lovely wooden fan-back chair of recent design. The seat was so large that she was able to stretch out to her fullest and do a rollover. Across the water, the red neon sign spelled out the letters HA FRUEN. One letter was missing. It should have been HAVFRUEN—"The Mermaid" in Danish—the name of a fish restaurant with French aspirations.

The theater boat just below was dark. The company was doing

only a morning children's show. The cabaret boat adjacent to the theater boat still showed some signs of life. A few people were heading home after a late supper in Nyhavn or after having been disgorged from the ferry from Malmoe.

A century and a quarter ago, Hans Christian Andersen had sat in the same position and looked out the same window. A picture on the apartment wall showed this. Lucas Fysst thought of his search for an unpublished story. In his mind, it had now become The Search. He got up, went to the kitchen and cut himself the tiniest corner of frugtfloedetaerte and poured himself a tumbler of Cherry Kijafa. This was rather more than his usual nightcap. He soon fell asleep in his desk chair, while Thomas Gray sat in hers and kept watch.

Unexpectedly, as though the bulb had burned out, the desk lamp went off. The street lamps below and the red mermaid neon sign cast pink and blue lights on the wall. The large quartz wall clock stopped ticking. From between the panes of the double-glazed window near the desk, a gray and blue nebulosity appeared. Vague at first, then taking shape, it headed toward the fan-back chair. Thomas Gray jumped off and resettled on the couch.

"I had a feeling you'd show up, H.C."

"Last night it was impossible for me to come, Froeken Graa, and tonight your friend Fysst stayed awake so long that I had to cool my heels, so to speak, outside in the misty air. Thank God for Kijafa. Thank God also for your invitation. Greatly appreciated. We ghosts are often struck from invitation lists."

"H.C., if you please. It certainly is misty. And mist does not work wonders for my arthritis," said Thomas Gray, as she rubbed one paw over the other.

"Try chewing a bit of comfrey leaf."

"Old wives' remedies. But tell me, H.C., when you do show up, you people don't really need the swirling mist, the shadows, the sudden drops in temperature, the rattling of chains, do you, in order to materialize?"

"Of course not. That's pure folklore," answered the ghost of Hans Christian Andersen. "Not necessary at all. Cheap theatricality. Nor is it necessary for ghosts to call

attention to themselves by playing the mandolin or shaking the tambourine. By the bye, how's Himself enjoying my old flat?" ("Himself" referred to Lucas Fysst.)

"Wonderfully well. This flat is quite central to Himself's activities. But he's worried, H.C. He's very much on the side of the animals when he allows himself to think about such matters . . . which is not very often. He's usually lost in manuscripts, documents, texts, books, palimpsests, scrolls, photographs of papyri."

"Interesting. Interesting."

"And just where do you stand, H.C.? I'm greatly disappointed in you. I sense ambiguity, profound ambiguity."

"On what issue?" the ghost responded after a time.

"To put it bluntly, on the whole issue of cats. Are you pro- or anti-cat? Basically, deep down, do you believe that cats have souls?"

A deep groan from the Victorian past emerged from the spectre. "I am certainly anti-dog. Early childhood experiences, you know."

"Very appropriate position to take: anti-dog. I've also suffered. But what about cats?"

"I vacillate, Fru Graa. My opinions are not steady. Now consider yourself, Fru Graa. You seem to have the power to call me up from the misty deep. I think you may not be unique in that regard. This is the hallmark either of the damned or of the metasouled. Which are you?"

"I do not like *ad feminam* arguments. I will admit that we cats, if we prepare ourselves properly, have certain powers or entitlements going back to ancient days that are not commonly available to other animals or to people. This does not mean that we cats have lost our souls because of it. History does not record that a cat bargained away her soul in Faustlike fashion (and therewith the souls of her descendants unto the ten thousandth generation) for the powers of communion with the Land of the Past or of its inhabitants.

"This power is our genetic inheritance, gained painfully over the millennia, and if it baffles you who are on the shady side of the bourne of life and who should now be party to the complete *mysteriae mundi*, then let us simply conclude that there are more things in heaven and earth, H.C., than are dreamt of in rational philosophy applied in either place."

The ghost of H. C. Andersen was a bit upset by this counterattack and responded initially by taking its handkerchief out of its frock coat pocket and giving its nose a tremendous blow.

Then suddenly, the ghost of H. C. Andersen swirled off the fanback chair and moved to the site of its old desk, where Lucas Fysst was snoring quietly. The ghost's spectral extent completely enveloped Lucas' body, as in a photographic overlay. It adopted a thinking posture—hand on cheek.

Thomas Gray looked on in fascination. I know, she said to herself, that one body can't be in two places at the same time, but apparently two bodies can be in one and the same place at the same time. Good to know. This principle might come in useful.

After its sudden change of venue, the ghost of H. C. Andersen resumed its former location.

"I am no philosopher, Fru Graa. A poet, perhaps, though largely failed, I'm afraid. A few stanzas still popular, but most of it washed away as the romantic and the sentimental failed increasingly to satisfy contemporary tastes.

"I may have misapprehended your position, Fru Graa. Perhaps I have been a bit anti-cat in my stories. I will say this in my defence: I did once write a story in which an intelligent cat played a role. I won't say that the cat was the hero of the story. No. Not at all.

However, given your sensibilities, I'm sure you would find in it no grounds for complaint. The story was never published. You'll not find it in my *Collected Works*."

"This changes matters rather a bit," said Thomas Gray looking relieved. "The manuscript still exists?"

"I believe it does."

"Where is it?"

"I couldn't say. Lost in Copenhagen, I suppose. I know only that it exists, but I have no way of telling where it is or how it can be recovered."

"Ah, then you are a pure existentialist, I judge. I mean in the mathematical sense, not in the literary sense."

"I do not catch your meaning, Fru Graa. I would say that I am now a pure nonexistentialist."

"Whatever. Well, that's very interesting to know. Not to worry. I'm rather an expert in scratching out and dealing with old manuscripts. There's a challenge here and I'll get to work on it. You'll come back after a bit, won't you? I'll give you a progress report."

"Ja. Jeg vil komme tilbakke. Farvel."

"Farvel, H.C."

The quartz clock resumed its ticking. The desk lamp was on again.

Lucas Fysst woke up suddenly and shivered. He shook his head to clear it. Then he said out loud (not yet realizing that Thomas Gray was sitting in the opposite chair), "Why, I've just had the strangest dream. I've dreamt a whole story about a Danish pastry. I think I'll write it down so as not to forget it. Perhaps it means something."

Thomas Gray smiled from whisker to whisker. Perhaps it does, she said inwardly.

11

Lucas Fysst's Dream: The Wienerbroed and the Cat

Many years ago, when King Frederik VIII sat on the throne of Denmark, there was a wienerbroed who lived in Copenhagen. The broed was reasonably round, reasonably plump, reasonably sweet, and had a reasonably juicy filling; in short, he was everything that could be expected of a respectable wienerbroed.

But the wienerbroed was unhappy, deeply unhappy. And this was the reason: There he was, you know, living in the pastry shop among the tarts and the tortes and the squares and the slices and the cakes and the balls and the crescents, all of which had proper Danish names. He knew that he had been born right there in the shop

on Frederiksborg Gade, but only he among all the pastries had a name that suggested he was a foreigner. This disturbed him greatly, and he decided that he wanted to be called merely a Danish pastry.

So the wienerbroed rolled out of his display case and went to the Ministry of Baking, not too far away, where such matters are sorted out. The Minister of Baking listened to the wienerbroed's complaint and then said, "Your problem is difficult and delicate indeed—foreign policy question and all that. But I'm afraid that I'm too busy with questions involving the national baking policy. I suggest you see the Underminister for Pastry."

So the wienerbroed rolled down the hall to where the Underminister for Pastry sat. The Underminister, who was as seasoned a bureaucrat as the minister, listened to the complaint of the wienerbroed.

"I humbly request and petition that I be allowed to call myself a Danish pastry."

"My dear sir. *You* may call yourself whatever you like, but it would be forbidden for the shop to call you a Danish pastry when you are a wienerbroed."

"Oh," said the wienerbroed, and he felt very sad.

"You see," explained the Underminister, "the matter is this. *All the pastries in your shop are Danish pastries.* They certainly are. But Regulation 19-D of the Code for Accuracy in Pastry Designations, of January 13, 1837, forbids a general name to be applied to a specific product."

"There is no way out for me then?"

"Well, I wouldn't say that. When it comes to regulations, there may be a loophole. Yes. Wherever there is a regulation, there is a loophole. That is a law of nature. I suggest you go upstairs—will you

be able to manage the stairs?——to where the Chief Counsel for Loopholes sits. Put the matter to him."

The wienerbroed rolled upstairs, found the Office of Loopholes, and explained the whole matter to the Chief Counsel. Now the Chief Counsel had much legal training in loopholes and prided himself on his knowledge and experience. He listened to the case and arrived at a decision.

"No. I am afraid I cannot see a way through to your request. You see, in other bakeshops there are French twists and spandauers and even English scones. These are names that reveal foreign ancestry in the deep past. I know this for a fact because my wife brings them back from Oster Gade. What if all these pastries came to me with a request to be called Danish? We should have chaos and confusion. A pretty kettle of fish indeed!"

So with downcast crust, the wienerbroed made his way back to the shop on Frederiksborg Gade. En route, he met a cat.

"Why do you have such a downcast crust, old thing?" asked the cat.

And the wienerbroed explained the whole matter to the cat.

"How old did you say you were?" asked the cat. "One day, did you say? You are suffering from a premature mid-life crisis, a common enough complaint. It comes in many different styles and flavors. Luckily, in your case, the cure is not too far away. I suggest you roll down to the harbour and onto the next boat to America."

And that is what the wienerbroed did. A few days later, he landed in New York and found himself on Broadway. It was not long before he saw an American bakeshop and, thinking he would be quite comfortable in that milieu, he went in the door.

"What have we here?" said the baker to his wife.

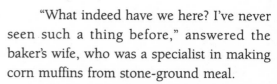

"What indeed have we here? I've never seen such a thing before," answered the baker's wife, who was a specialist in making corn muffins from stone-ground meal.

"Oh," said the wienerbroed. "I'm a Danish pastry. That's what I am. I may be a bit stale from my journey, but I'm still plump and sweet and juicy inside."

"Of course you are," said the baker, putting him into the case. "You're a Danish pastry, and you will be the first of many in New York."

And so, you see, if it hadn't been for the sharp analysis and quick action of the cat, the New World would never have known the pleasures of Danish pastry.

Of course, the sad part is that for reasons you can well imagine, the wienerbroed never returned to Copenhagen.

And the cat? Well, she had a lively career as Consultant to the Department of Loopholes, and the Chief Counsel's wife got a green silk dress to dance in.

12

Instructions Are Received

Lucas Fysst, with Thomas Gray tucked securely into a canvas shopping bag that he had bought at the IRMA supermarket, traveled northwest on the S-tog (the intercity rail line). They had an appointment with a certain Charlotte Kochloeffel, who lived on Taastrupvej in Husum, a suburb of Copenhagen.

The appointment came about in the following way. One of the portraits hanging on the wall in Nyhavn 18 was that of a lady. During the conference and subsequent research, Lucas had simply forgotten about the photograph. One afternoon, however, by performing furious antics on the couch just below the portrait, Thomas managed to remind him of it. He examined it carefully. There was a little brass plaque attached to the photograph that read:

> Dorothea Lillyveld (1823–1885), and her husband, Carl Lillyveld, were friends of H. C. Andersen in the last decade of his life. They took care of Andersen while he lived in this flat.

It struck Lucas that Lillyveld was not a usual Danish surname. If the family survived at all in Copenhagen, there would probably not be hundreds or thousands of Lillyvelds listed in the phone book, as is the case with Jensens and Petersens. This conjecture proved accurate. After a few calls and a few referrals, Lucas was put on to a Charlotte Kochloeffel, who appeared to be some sort of descendant of Dorothea Lillyveld and who seemed quite willing to be interviewed. Perhaps, Lucas thought, there was some family tradition about unpublished manuscripts. It was a lead worth following up.

Taastrupvej was a respectable street, well parked with cars, and the house Lucas sought was a modest two-story affair, one of fifteen or so attached in a long row. There was a playground opposite where towheaded children played on swings and slides. A plastic sign on the door read:

Charlotte Kochloeffel

Kropsmassage

The door opened, and a slight woman of about fifty appeared. She was a blazing redhead (rather bouffant, thought Thomas Gray,

INSTRUCTIONS ARE RECEIVED   83

who peeked out of her bag), and she wore a yellow sweatshirt over spandex stirrup pants with bright blue vertical stripes. She was barefooted. Danes often remove their shoes in their homes.

"Ah," said the woman, "you seem decent. Would you like to see the front room? It's nice and sunny. You share the W.C. with me. But no telephone. And no TV. I won't allow them. Outside energy impures my personal channels."

Lucas looked confused.

"Come in, come in. If you like the room we'll sign the papers."

Still confused, Lucas was led into a small kitchen. He sat down at the kitchen table and set his shopping bag on it.

"Excuse me, but can you tell me if you are a direct descendant of Dorothea Lillyveld?" Lucas asked.

"Certainly. I'm not. I am a direct descendant of Dorothea Lillyveld's brother, Marcus Gylden."

"Through some female line, obviously."

"Right. Marcus Gylden's daughter Clara was my father's great-great-grandmother. But really now, what has all this to do with renting my room? New regulations from the Ministry of Housing? It's come to that, has it?"

Soon, Thomas Gray thought, in my country at least, blood types will be required on all official documents.

Thomas popped out of the shopping bag onto the table and stretched.

Charlotte looked at Thomas. Now was her moment of surprise.

"This is not my cat! My cat is orange."

"No indeed, Fru Kochloeffel, this is not your cat."

"Where is my cat? Oh, you aren't from the veterinary clinic? Well, do you come for a massage? Do you want to rent my front room? What are you here for?"

"Madam, I rang up yesterday about possible recollections of Hans Christian Andersen in your family. Unpublished manuscripts. That sort of thing."

"If you say so. I must have been in alpha mode when you called. My English understanding is not good when I'm in alpha. Excuse me, please, for one minute. I must keep to my schedule."

Charlotte swung into the adjoining living room and put an exercise tape on her player. She began a regimen of calisthenics and contortions of a very nonstandard kind. Loud music diffused into the kitchen.

"You are a churchman? I hope what I do now does not shock you."

"Nothing human is alien to me," said Lucas.

The question, thought Thomas Gray, is whether what we are

witnessing is human. I would consider it more in the amphibian or herpetologic line.

"Well, I'm now back in gamma mode, so just ask what you want," called Fru Kochloeffel from the living room. "Maybe I'll know some answers." Her voice barely penetrated the music.

"I understand that the Lillyvelds and the Gyldens were great friends of Hans Christian Andersen toward the end of his life. It's in his biography."

"He used to visit them regularly. Stay with them. Eat with them. They were very rich. He was very famous. Dorothea kept, how you call it, a literary salon. He died in her house. I remember my father telling me that she looked after Andersen for years."

"Are there any Andersen anecdotes in your family?"

"Yes. He hated the fruit soup. There must have been lots of stories, but they've been watered down."

Most stories get fattened up over the years, thought Thomas Gray.

Charlotte went on. "He was a strange man, that Andersen. A cranky, melancholy bachelor. Civilized but irritable. A tall and ugly man. But apparently they got along with him. I heard he used to come and read his stories to them. And then he would sit and make paper cutouts, one after another. I think my father inherited some of them."

"Did you ever hear that Andersen gave any of his original manuscripts to Dorothea Lillyveld?"

"He could have done so. I heard that he left behind much letters to them and also sheets and sheets of diary writing. Dorothea Lillyveld pulled the material together after he died. You know, I am a writer, too."

"Really? Good for you! What do you write?"

"Letters of complaint, mostly. To the *Berlingske Tidende*, the big newspaper in Copenhagen."

"Well, that's a start. Does your family have any original Andersen material?"

"Not that I know of. I'm not that close to my family. He was a famous man. If the family had that kind of thing, they probably sold it or gave it away years ago. Am I rich? You see I am not."

"Are there any anecdotes in your family that tell whether Andersen liked cats? His stories are full of animals."

"Say your question again. The music is loud."

"Do you know whether Andersen liked cats?"

"Please?"

"Did Andersen like cats?"

"How would I know such a thing?" Fru Kochloeffel responded. She turned the tape off and came back to the kitchen.

"Thank you very much. You have been most kind. Here are my address and phone number should you think of something."

"Yes. Maybe I shall ask my brother. Oooh. Nyhavn. What a very fancy address!"

Thomas Gray popped herself back into the canvas shopping bag.

"My cat Hugle rides on my cycle. You certain you don't want an appointment for a massage? Or to rent my front room? You can have one shelf in the refrigerator. Well, farvel."

"All three would be nice, I'm sure," Lucas Fysst said, and he left quickly before the alpha mode had a chance to set back in.

. . .

Back in Nyhavn 18, Lucas Fysst mused out loud. "Even though we've had no luck whatever, Thomas, I feel even more strongly now than before that there is an Andersen manuscript that has not been published."

Thomas nodded assent. She had just finished off a bowl of Whiskas (she had found the food at the Black Raven awash in sauce and much too rich for her), and she was preparing to watch a German soap opera on Danish cable.

"And I think further, old girl, that the manuscript has a decent cat in it. Not a spectacular cat—not a cat in your bracket, old girl—but a cat of whom one should not be ashamed. A cat sufficient to rehabilitate the reputation of a great man. And Andersen was a great and humane man. We need not, repeat not, suffer any revisionist criticism of his oeuvre. There must be other ways of making a splash in the learned world than by turning to revisionism."

Thomas nodded assent.

"And I think further that the manuscript has been lost in Copenhagen and therefore is to be found in Copenhagen."

Thomas nodded assent.

"The manuscript is not in this flat. Grant me that. This flat was totally remodeled not so many years ago and completely refurnished. Nothing hidden in legs of tables. No point in looking for secret drawers or gaps behind wall panels, et cetera."

Thomas agreed. No point whatever in et cetera.

Ten minutes later, the phone rang.

"Fysst here."

A man at the other end of the line spoke English with an accent that sounded Central European to Lucas, probably Slavic. The articles "the" and "a" were omitted. The "l" was soft, almost a "w."

"I believe you are in market for original Andersen manuscripts?"

"Why yes, after a manner of speaking. Speaking of speaking, to whom am I speaking, please?"

"I prefer not to say for time being. I assume you are prepared to pay price?"

"Why, my dear sir, I never thought of the matter in terms of price. After all, we're dealing with Danish national treasures. Such things are beyond price."

"Not to be too philosophical, Dr. Fysst. Everything has price."

"Well, er, er, . . . yes, well, hmmm. I suppose I might come up with a few pounds. It'd be well worth it."

"If you want it and you pay for it, then it's worth it. That's what economics means."

"Yes. I suppose so. I'm no economist. Let us say, then, that I might come up with a few more pounds, if what you have is what I want. Naturally, I shall have to examine what you have before paying you. What is your price?"

"Five thousand pounds sterling in hundred-pound notes. Quite reasonable for national treasure, don't you think? This call is now too long. I put instructions in mail. Goodbye."

"Goodbye, Mr. . . . ? Well, Thomas, what do you think? Should we proceed?"

Thomas Gray looked slightly dubious at first. Later, when she had thought the matter through a bit, she gave a nod of assent.

"I'm glad you agree we should proceed. Nothing ventured, nothing gained, what, Thomas? It's distressing that we have to deal with characters who live on the shady side of the street, but those are the realities. When our story has worked its way to completion, the newspapers will stress the shady, call it romantic, and ignore the scholarship."

After a few days, a letter containing instructions and postmarked Copenhagen was delivered to Nyhavn 18. It had no return address. The enclosed note, typed and unsigned, read as follows:

> Tuesday week (date given), go to Viking Ship Museum, Roskilde. At 14:15 exactly, not before, walk into coffee shop. Many tourists there but will be table where no one sits. Umbrella and raincoat on chair and brown envelope on table. Examine contents for ten minutes. If you like, remove contents and insert five thousand pounds as agreed. Leave on table. Not follow or try discover or identify yours truly.

"Well, the seeds I've planted have sprouted," said Lucas Fysst to Thomas. "Strange fruit. I suggest we keep this date. It may turn up something. One can never tell."

Thomas agreed strongly.

"On the other hand, there's no reason for anyone to treat our legitimate search in a shady manner. It should be amusing, though. And you know, if we're lucky, we can combine the trip with a visit to Mr. Helgerssen. It's in the same direction. He's the rich collector of rare books. *Very* rich, I'm told."

Lucas rang up Helger Helgerssen's secretary. He learned that a week from Tuesday was not possible. Mr. Helgerssen would be in Milan. The following Friday, Saturday, and Sunday were possibilities, though. Lucas set the appointment for Friday, because trains and buses ran more frequently during the week.

A half hour later the secretary called back. "I rang Mr. Helgerssen in Milan. He said he recognized your name. Friday week would be agreeable. Mr. Helgerssen invites you for lunch. At 1:00 P.M, shall we say?"

"Splendid. Incidentally, I shall be coming with my associate, Thomas Gray. She's a cat."

"I understand, Mr. Fysst," said the secretary, who did not understand at all. "Then there's no need to lay another place."

Thomas Gray did not enjoy the slight. Every need, indeed, to lay another place, she thought.

The secretary continued, "I'll pick you up at the station in Nykoebing. That's the Nykoebing in Sjaelland, mind, not the one in Falster."

13

The Plain Brown Envelope

In response to the phone call offering a manuscript for five thousand pounds, Dr. Fysst set out for Roskilde and took along his associate Thomas Gray in the canvas shopping bag. The local train sped west from Copenhagen through village after village: Roedovre, Albertslund, Hoeje Taastrup, Hedehusene, Trekroner.

Lucas looked out the window. The train passed a private home with a goat in the yard. Denmark had turned out, he thought, to be a very comfortable country, even though he could not understand the spoken language. The people looked pretty much as they did in England and acted pretty much as they did in England. Why should it not be so? Well, wasn't a good part of his country Danish at one point in its history? Weren't half the place names in eastern England Danish in origin? Wasn't the land administered locally according to Daneloven, the traditional Danish law? Wasn't even William the Conqueror a Norseman? Wasn't it possible that he, Lucas, had some Danish blood pulsing in him?

Lucas recalled that a few days back, on a slow morning at the IRMA supermarket checkout counter, a clerk had given him his own assessment of Denmark. "We are not poor and we are not rich. We pay very high taxes. Higher than you do. Then certain important things are free. On the whole, people are satisfied. Happy? Who knows what happy means? We're reasonably comfortable. Complaints? Certainly, but not too many. The social arrangements seem sensible. A few people find loopholes—ways to pile up fortunes. It's not too bad, though. We're a small country. We want our independence; we want to be ourselves. We worry about the large countries of Europe."

On the left side of the train they passed a bank of windmills, each with a sleek, tapered shaft on which three aerodynamic blades were mounted. Starting out from a city whose skyline consists of baroque steeples, it was only too easy to forget that Denmark is a modern industrialized country with a modern government and modern problems. Denmark is not a country that can be summed up by the words beer, cheese, sausage, and pastry, much as Lucas enjoyed these things.

The windmills underlined this view. The rotation of their blades was slow and dignified. A conference participant, Professor Lars Aunstrup from Aalborg, had told Lucas that he owned a windmill. On questioning, he modified the statement and said that he owned one twenty-fifth of a windmill. His mill generated enough power for about twenty-five private residences. A company erects the mills and then sells shares. After about ten years, so the calculation goes, the mortgage is paid off and thereafter the power is free.

A half-hour trip brought the train to Roskilde. At the station they taxied to the Viking Ship Museum at the head of the Roskilde Fjord—a five-minute ride. Lucas released Thomas from the shopping bag and they went inside together. As soon as Thomas Gray saw the restored ships, she jumped into the nearest one. Her tail beat in a violent and slashing staccato. She ran up to the high prow.

She ran up to the equally high stern. She wailed. She paused and looked toward Lucas, who was standing on the other side of the guard ropes.

"Now it comes out, old girl," Lucas said to her. "You're really of Viking origin. Your genes seem to be vibrating atavistically. They find the environment familiar and you run back and forth, berserk. Did you know that 'berserk' is a Danish word describing the battle state of the Viking warriors? 'Drunk with slaughter' is what it means. Did you know that the monks and nuns on the east coast of Britain inserted the words *a furore Normannorum libere nos* into their daily prayers? 'Free us from the fury of the Norse.' How could you know? Calm yourself, old girl. Pretend you're civilized."

Thomas was still greatly agitated. "Don't translate," she thought. "Don't translate; it irritates me," and she continued her tail slashing. Then she got a grip on herself and realized that she was almost in the twenty-first century and not the ninth. Her sympathetic fury abated as the genetic memories of ancient battles were diluted by the sight of small children trekking through the museum in groups, eating ice cream.

My identity is torn to bits, reflected Thomas Gray, I am a confused package. I used to think I was of Dutch origin, but visiting these ships encourages me to trace my family roots deeper. It's conceivable that my ancestors went to Holland on one of them. The Vikings sailed up the mouth of the Rhine and raided right and left. On the other hand, the Swami thinks my soul is that of his guru.

Lucas was thinking his own private thoughts. Listen, said the Beowulf-bard opening his great epic, and I shall sing you of the glorious and daring deeds of the Danish kings. When will glorious deeds be other than slaughter? Is such a transformation possible while the memory of the past remains a vital part of the present consciousness? The millions of kroner spent on exhuming these ships from the mud of the fjord and encasing them in a building of splendid proportions and of costly construction is an act of renewal, a rebirth of national pride, a statement of vital energies transduced. Now answer this: When does a knowledge of history lead to jingo? When does its lack lead to meaningless existence? When should memory speak out, and when should memory obliterate itself?

At any rate, it gives hope for the world that the cruel Danes who held sway over these northern lands and seas for so long have been tamed by time, reshaped and smoothed into the most agreeable and most sensible of people. If it happened here it could happen again with the nations that are now roaring so loudly. Such a thing does not contravene the laws of physics. It would spell the end of history as the chronicle of kings and battles and greed and revenge and begin a new book: the record of the refinement of the soul.

At 14:15 precisely, Lucas and Thomas entered the coffee shop. The shopping bag Lucas carried was now empty. There was no money in it; there were no five thousand pounds in sterling, kroner, or dollars. If the manuscript was in any way significant, payment could be arranged for later. There was no need for rushing.

The little lunchroom was quite full of tourists. By the large picture window that overlooked the Roskilde Fjord, one table was empty. Lucas spotted the raincoat and umbrella on the chair seat and a plain brown envelope on the table. He did not try to identify which of the many diners might have placed it there. He went over and sat down. He extracted a manuscript from the envelope. It consisted of a sheaf of sixteen sheets of old paper, scribbled on with an old-fashioned pen in old-fashioned Danish cursive. It was Andersen's story "The Elf Hill." It bore the stamp of the Andersen Museum in Odense.

Lucas examined the sheets one by one, carefully and professionally. Time was passing. He put his hand to his chin and thought a bit. He spoke to Thomas Gray, who was on the opposite seat. She leaped onto the table. She looked at the manuscript, pawed it, and smelled it. Then, with a brisk motion of her paw, she knocked it to the ground. Lucas picked up the manuscript, put it carefully into its envelope, and left it on the table. The two of them then strode out of the museum and walked back to the railway station.

The train back to Copenhagen was not crowded, and Lucas and Thomas had four seats to themselves, two and two, facing each other.

"Even though we left the manuscript on the lunchroom table, I should like to tell you more or less what was in it," said Lucas. Thomas expressed a willingness to listen, and Lucas began.

"'The Elf Hill' is really one of Andersen's finest stories," he said. "Quite cheerful. Rather the exception, you know, and it gives a very good feeling for what is human in the animal.

"The lizards noticed that there were great goings-on at Elf Hill. More than just a cocktail party, it seemed. Lots of cooking, the silver plate polished up, the fancy ladies practicing their dance steps . . . that sort of thing. Obviously the King of the Elves was laying on something special.

"The lizards were not privy to the invitation list, and even the earthworm who had his ear constantly to the ground didn't know too much. Ultimately the little old lady who keeps house for the Elf King spilled the beans to the night raven.

"For the grand ball, there was an open invitation to the whole world—to humans even, especially those who talked in their sleep. But invitations to the banquet were another matter entirely: very exclusive; very posh. The merfolk were invited and the will-o'-the-wisps, and the various churchyard irregulars. Disgusting creatures, those fellows, really!"

"Who are the churchyard irregulars?" asked Thomas, who made a hobby of keeping track of the irregular. "Something like the Baker Street Irregulars?"

"Not at all. For starters, the death-horse and the pig buried in the churchyard. The horse comes out every night and, managing on three legs, clops over to the houses of those who are about to die. And then there are the type-one demons, *id est,* those who qualify by possessing tails. And then the church dwarf. However, I must inform you that despite certain loathsome invitees, the banquet was, in some sense, so exclusive that the ghosts were not invited.

"What? No ghosts? Pity. I've met some rather nice fellows in that category," said Thomas.

"What you tell me isn't surprising. Cats have always had a communication channel to the transcendental. Or so it is said. But to get on. The kitchen, as I have suggested, was bubbling with activity. Frogs were roasted, snail skins prepared, salads of hemlock and mouse noses were tossed, barrels of beer made by the marshwomen were laid down. Very solid food indeed. And for dessert there were rusty nails and broken glass from church windows—coloured glass, I believe—very special, very expensive.

"The guests of honor were the Great Goblin of Norway and his two sons. And there were also a few assorted Norwegian trolls and goblins of lesser rank who came as part of the entourage. The father was a rather nice chap; the sons . . . hmm, the less said about their manners the better. Well, I'll let you hear one piece of rudeness. The Danish elves had opened up their hill quite nicely. Everything was spiffed up, polished up, the curtains were fastened neatly with snail slime.

"The boys took one look at all this and said, 'Is that what they call a hill in Denmark? In Norway, we'd call it a hole.'"

Thomas Gray interrupted. "That's an old joke. Right out of *Joe Miller's Jest Book*, first edition."

"Exactly: published in 1739, I believe. And here's a more recent version: The famous French painter Henri de Toulouse-Lautrec was extremely short, you know. Once, when he stood for a group photograph with some other painters, an onlooker called out, 'Who's that man standing in a hole?'

"But to carry on: The Great Goblin of Norway had decided it was time for his boys to be married, and a selection was to be made from among the seven daughters of the Elf King. As I said, the Great Goblin was a nice sort and quite liberal. In those days it was uncommon to have a selection committee of the type now in university affairs. One after another, the girls would show off their accomplishments, and the boys would select whomever they liked.

"The first daughter put a piece of wood in her mouth and disappeared. What a trick! But neither the father nor the sons saw much virtue in that. The second daughter had a shadow, and she walked alongside it. Everyone knows, of course, that elves and goblins have no shadows, so this put the poor thing in an anomalous position.

"The third girl knew how to brew at the marsh and how to lard trees with glow-worms: impressive and useful, what? The fourth owned a magic harp which, when played, compelled listeners to do as she wished: very dangerous, very unstable. The fifth's sole accomplishment seemed to be that she wished to marry and settle in Norway: Good enough when considered by itself, but the real reason was not for love of the Norwegian goblins as much as to escape to Norway's mountains from the flood that was surely coming.

"What would be called a *mariage de convenance*," said Thomas Gray, who could express herself in a kind of feline-French much abhorred by the Academy in Paris.

"Precisely. The sixth daughter spent all her time sewing her shroud—a pious gesture but not conducive to mirth and cheerful thoughts. Finally, the seventh daughter could tell stories without ever repeating herself."

"A veritable Scheherazade," put in Thomas Gray.

"Just so."

"And the upshot was?"

"And the upshot was—not to elaborate on a synopsis that's already much too long—the upshot was that the boys, who were both rude and immature and preferred to drink, pontificate, and sleep, decided to carry on with their bachelor lives. Their father, who was a widower, married the Sheherazade and experienced much joy in his selection."

"My feeling is," opined Thomas Gray, "that what we have here is a storyteller's vision of a marriage made in heaven."

"There is no doubt of it."

14

The Kongekat's Story

When they got back to the flat in Copenhagen, and as a result of the fiasco at the Viking Ship Museum, Lucas sat down at his desk and dashed off a letter to a man he had known for a number of years.

Nyhavn 18
Koebenhavn

Dear Dr. Kalamos:
Your Danish representative found the wrong customer. Your facsimile of "The Elf Hill" manuscript is very good indeed. I congratulate you. It would have taken me some time to

identify the manuscript as coming from your atelier, certainly more than the ten minutes your representative allotted me for examination. But Thomas Gray did it immediately. She has her methods, you know.

Now here's the thing. I don't want a manuscript that was lifted from the Museum at Odense, not even a facsimile thereof. I'm looking for a genuinely new Andersen manuscript that is unknown to the general public and has remained unpublished. If you know of such, write me at the above address. I'll be there for awhile yet.

*Yours faithfully,
Fysst*

Within a few days, he received an answer.

*1345 B Mariahilfer Str.
A-1040 Wien*

*Dear Dr. Fysst:
I beg you a thousand pardons. It was sheer stupidity, of course. My salesman is new to the business. He simply did not know your professional status. But tell me, what imperfections did you or your cat find? Such knowledge will help me in improvement of future products.*

Regarding your request, I do not know of any such manuscript as you look for, but I shall make deep inquiries into your behalf. I will write you only if something positive.

*Yours truly,
M. A. Kalamos, Ph.D.*

The Kongecat's Story

And so The Search went on. Even as Lucas Fysst was making further inquiries among the book dealers of Europe and the libraries of Denmark and Dr. Kalamos was proceeding in his own nonstandard way, Thomas Gray was playing the tourist. For a change, the day was sunny. Research could be left for gray days or for the evenings. She found herself in Amalienborg Plads ten minutes away from her residence in Fyrskib Nr. XVII and very near the Royal Palace. Four rococo structures surround an equestrian statue of Frederik V, a rather unimportant king of Denmark, whose virtues are spelled out in Latin in letters that have become unreadable. Just as well, Thomas Gray thought.

The Royal Life Guards, magnificently clad, bayonetted rifles on their shoulders, swords at their waists, shakos on their heads, paced slowly, singly and doubly, to and fro. Crowds of tourists piled out of their buses and snapped pictures. Mogens Hvidvin, who had set up the conference and was Lucas Fysst's Copenhagen connection, had told him that the Royal Life Guards were now maintained merely for the tourists. They bring in business.

Within the central chained area surrounding King Frederik, Thomas Gray encountered a sister cat, who was gray-yellow and slightly misty. She looked familiar.

"I remember you. You are Mewsli," said Thomas.

"And you are the Englisher, not?"

"Not. I mean jot. No, not jot. I mean yes. Now that I've looked around Copenhagen a bit, I have the impression that all the kings of Denmark are named either Christian or Frederik."

"That has been true for many hundreds of years. It makes things simple for us."

"It leaves me in a horrible confusion."

"How's your consulting going?" asked Mewsli.

"Poorly."

"But you remain?"

"A job is a job. One does not give up easily. Good things often take time."

"Tell me."

Thomas Gray described the job in the briefest manner.

Mewsli said, "I know all about Andersen. He wrote about trolls. I know all about trolls. Besides, I may be able to help."

"How so?"

"I'm the Kongekat!"

"The what?" asked Thomas Gray.

"The Kongekat. The Royal Cat."

"Well, well. Should I bow? Should I walk backwards upon exit?"

"Bow? No. But proper respect would be in order."

"How long have you had the job?" Thomas Gray inquired.

"Two years, give or take a few weeks."

"Do you get around much? Are you confined in any way?"

"In a few days I shall have a litter."

"That's clear, but it's not what I meant. Are you free to go where you want? Or do you have to put up with bodyguards and all the other royal paraphernalia?"

"Not at all. I'm totally unconfined. Denmark is democratic to a fault. Let me ask you a question now. Are you the Thomas Gray who was in the Sunday papers a while back?"

"The same," Thomas Gray answered, feigning false modesty.

"The Queen heard that you and your assistant were in town. She keeps up with that sort of thing. She spoke of you flatteringly. Said you were part of the learned world."

"I am that. But spare me the sin of pride."

"We Danes could use a bit more pride. Now on to business for a moment. I'll set up an audience with the Queen for you and your assistant. Then you can ask about the Royal Accessibilia."

"What's that?"

"Collections. Objets de virtu. Old books. Manuscripts. Art. Medals. Jewels. All of it is the personal property of The Family. Accessible only to the Royals. The Andersen manuscript you look for may have devolved to the glass cases in the Royal Palace."

"Yes, do that. I'd be greatly obliged for an audience. If I'm not too curious, how do you like your job?"

"Many perks."

"Such as?"

"Travel. Summer lodges. Royal Yacht on the Mediterranean. Paid maternity leave."

"Duties?"

"Light. Some mousing, but mostly social."

"Are you unionized? I hear that unions are strong in Denmark."

"Lady, I'm the Kongekat. I'm a union of one."

"Touché. Incidentally, how'd you get your job?"

"Do you have an hour to spare? It's a story I love to tell."

"Tell away. I have all day."

"I was born," said the Kongekat, "in Skoerping, a small village in northern Jylland, near the Roldskov, the Forest of Rold. In this village, quite modern in all essentials, one still sees a few horse-drawn wagons loaded with farm produce, but they have become rare. Apart from the month of July, the most exciting event in Skoerping is the series of community soccer games. Once there was a shortage of leeks in the market and that caused tongues to wag for a few days.

"My forebears have lived in Skoerping for centuries. There is a tradition in my family that one of my ancestors in the female line was Chamber Cat to Queen Margrethe I. That was many years ago in what humans call the thirteenth century. The ruin of her castle can still be seen not too far from Skoerping.

"My mother worked in the kitchen of the Hotel Roldskov. After I left the litter, I spent many happy hours in the nearby forest, digging in the mosses and under the leaves.

"I would like you to know that the Forest of Rold is the largest forest in the whole of Denmark. It is really quite a special place, because it has groves where a certain kind of beech tree grows. The groves are called the Troldskov—the Trolls' Forest —and, unlike most beech trees, whose trunks and limbs grow straight, those of the Troldskov grow in a twisted and gnarled fashion. It's not that the trees are puny; they are not. It's not that they are diseased; they are not. It's simply that their limbs grow crossing one another in a curly, spirally way—and the Troldskov is the only place in the world that has such trees.

"These trees are said to be under the influence or tutelage of the trolls—*trolddomskraft* would be the technical term for it. And before I proceed with my story, I'd better say a few words about the trolls. They have gone into a partial eclipse as a result of the prominence of the Hobbits and other fictitious creatures. Now your Norwegian trolls are fierce indeed: giant, one-eyed monsters, relentless in their savagery. By way of contrast, Danish trolls, often called *nisse*, are very small, much more civilized, much more given to practical jokes and annoyances. They are even helpful on occasion. Under duress, they can be liars and thieves. The daylight is dangerous to them. If caught in the daylight, they turn into roots or stones and then require reconstituting. This being the case, they are most usually observed around dusk.

"All trolls have two left feet, and this has some advantages. One characteristic of the trolls that I find admirable is their hatred of loud noise. This is said to derive from a traumatic experience suffered centuries ago when the great god Thor, in exasperation, threw his hammer at them. The world has paid much attention to chemical pollution and not enough to sound pollution. I could continue ranting in this vein. It's one of my bêtes noires, but it would get me off my story.

"Occasionally—it doesn't happen with every tree—several branches of a beech tree may grow together, forming loops or eyes. Such beeches have magic powers. They relate themselves to diseases of the knuckles and the teeth, and it is said that a person who goes through one of these loops is immune from these diseases.

"I have been through these loops many times, and I'm happy to tell you that I've never experienced difficulties with my teeth or any of my bones. You may take this as evidence in favour of this kind of preventive medicine.

"One day—I must have been a year old—I was deep in the forest when I found a beech tree that quite remarkably had two loops. And I thought to myself that if one loop is magic, then two must be supermagic. And I hopped through them both, one after another. The darkness was coming on, but I was so absorbed in exploration and in finding a double loop that I hardly noticed it. But enough was enough and I turned to go home.

"On my way back to where I lived near the Hotel Roldskov, I spied a small brown rabbit in the distance. The rabbit saw me and hesitated. I pounced and was able to press the rabbit to the ground.

"'Let me go. Let me go,' said the rabbit in a high, squeaky voice.

"'How's that?' said I. 'You can talk? I thought you were a rabbit.'"

"'If you can believe I'm a rabbit,' squeaked the rabbit, 'then you can believe anything. Let me go.'

"'Then if you're not a rabbit, who are you?'"

"Rather pompously, the rabbit answered, 'I am Joern Mosemand, Secretary-Treasurer of the Roldskov Trolls. I am temporarily in the shape of a rabbit for diplomatic reasons and I don't care to discuss it with you. Let me go, I say.'

"'Do you live around here?'" I asked.

"'Quite close. I own an underground villa called Au Contraire. That's French for "In the country".'

"'Don't translate,' I said, 'I don't need it.'

"'Sorry. Now let me go. Let me go.'

"'I most certainly won't let you go,' said I. 'It's common knowledge that if one catches a troll, one may force him to grant one's wish. I will not let you go until you have granted me a wish.'

"'Let me go. Let me go, and your wish will be granted,' Joern Mosemand protested vigorously. 'What do you want? Fish? Fame? A life of frivolity?'

"'Don't be patronizing, or else I'll keep you here till mushrooms grow in your beard.'

"'What do you want, then?'

"'I don't know at the moment. One doesn't go through life carrying around a prewormed fish—sorry—a preformed wish in the expectation of catching a troll. At least I don't. You must grant me a wish on a contingency basis.'

"'I cannot do that. We trolls work to rules, and what you suggest would be against our rules. You must state your wish now. I will then have to verify that your wish lies within the category of acceptable wishes.'

"'Then I shall keep you pinned down forever.'

"'No. You mustn't do that. Must not, do you hear?'

"Though I had him pinned, the rabbit stamped the ground furiously with his forepaws. Then he relented.

"'I'm late already. Well . . . I'll make an exception this once. You may specify your wish at a later date. But only one wish, mind you.'

"'You trolls are unreliable. Swear.'

"'I swear.'

"Well, I had the promise of the Secretary-Treasurer of the Roldskov Trolls, obviously an individual of some responsibility. If he was not to be trusted, who was? So I took a chance and let the rabbit go. One must have a measure of faith, even with trolls. At the same time, one must realize that there is no ultimate guarantee of anything.

"As soon as I released the rabbit, right before my eyes the rabbit turned into a troll, and before I could catch my breath at the transformation, he disappeared into the darkness of the forest.

"I did not want to rush precipitously into making my wish. I thought about it for some while and then made my decision. The village of Skoerping, the area around the hotel and the troll groves had been my entire world. Although my life in this world promised to be entirely decent and have many unsung rewards, I conceived of the idea of going to Copenhagen to see something of the larger world before settling down to a life of rural tranquillity. French cats have a saying: There are two kinds of people in the world: those who are blessed and those who are spared. In hearing what followed, you must judge into which category I've fallen.

"I thought also that when I was in Copenhagen seeing the sights, I should pay my respects to the Queen. Why shouldn't I? After all, I had a certain connection there. I would tell the Queen that my remote ancestress had been Chamber Cat to Margrethe I and trust the present Queen Margrethe's sense of the fitness of things to get me through the interview. Anyway, to visit Copenhagen and to say hello to the Queen—nothing beyond that, mind you—seemed a modest enough request. One must never press the power of trolls to its breaking point. So I wished for that. And I waited.

"For some time nothing happened. If I thought that, as in 'Puss in Boots,' I would be miraculously transported to the Royal Palace, I was mistaken. Nothing of the sort occurred. I began to think that Joern Mosemand, the Secretary-Treasurer of the Roldskov Trolls, had gone back on his promise, and I wondered whether I could catch him again if I went back to the forest at dusk and went through the doubly looped beech tree. I also wondered whether there was some higher tribunal somewhere, in which I could bring suit against a troll for breach of promise.

"I asked around for opinions about trolls. I was informed that all Scandinavian trolls, ghouls, and malign elementals are descended from the monster Grendel, himself a misbegotten descendant of the sons of Cain, who was finally dispatched by the hero Beowulf. The Danish trolls, however, have evolved into a rather benign branch of this unpleasant family. They are politically conservative and ecologically radical. They advocate a strengthened monarchy and a restoration of the flora and fauna to the condition of the ninth century. They are quite pagan. Their position with respect to modernity is inconsistent. On the one hand, they are thought to play havoc on occasion with computer networks out of pure malice. On the other hand, they are said to display a fondness for frozen dinners. I found this personality profile interesting, but of course it was irrelevant to my case.

"It was now getting on toward July. And at this point, I must tell you something that you, as an English cat, may find both hard to believe and dull to listen to. In northern Jylland (where my village of Skoerping is located), the Fourth of July—the American Independence Day—is celebrated publicly.

"Why, you may ask. It goes back a hundred years or so, when a number of Danes, who had migrated to the United States and had made great business successes, returned and initiated the custom, and to this day, parties, speeches, and concerts are laid on. Dignitaries come from America and from Copenhagen.

"I had heard that a grand feast was planned to take place in a field not far from the Troldskov. A Royal Personage—but not the Queen herself—would be present.

"The Fourth of July came. The flags of the two nations were flying. A band played. Formal words were spoken in Danish and English (Americanese, to be precise). Many tables were set out, including a head table. At the center of the head table sat the Royal Personage. On her right was a Danish film star from Hollywood. On her left was the fiskehandlerforeningsdirektoer, a representative of the North Jylland Fish Industries, a commercial group that had socked some money into the celebration. I watched from a nearby tree.

"The waiters brought in the forretter—the starters. First came pickled herring, naturally. What else could it have been? After all, that is the national starter, if not by act of the Folketing, then by long tradition. The Royal Personage took up her fork and toyed with her appetizer. (I think I can reveal without serious breach of privacy that the Royal Personage was a lady.) From my perch in the tree I could see that she did not care for the herring. But remember, next to her sat the fiskehandlerforeningsdirektoer. For her not to have eaten the dish would have been the commercial equivalent of lèse majesté and national heresy. The film star and the fish man were now well into their herring. The Royal Personage still hesitated.

"But I did not. I jumped from the tree. I ran to the table. I sprang up and knocked the Royal Personage's starter to the ground. It all happened in the flash of an eye. The waiters were confused. They brought napkins to cover the soiled spot on the Royal Lap. They fussed. They offered more herring.

"'No, thank you,' said the Royal Personage, quite relieved. 'Let's get on with the banquet.'

"I ran back up to my limb. When the banquet was over, I came down, and putting on a look of the hungriest, lonesomest, most abandoned, most apologetic, most pliable, most intelligent cat in the whole universe, I rubbed against the Royal Ankles.

"The Royal Personage took me up in her hands and popped me into the large red and white handbag she was carrying. 'You were wonderful! Your talents are wasted here,' she said. 'There are many banquets in my life.' As she left the table, I poked my head out a bit

and caught sight of a brown rabbit running across the field toward the grove of twisted beeches. Joern Mosemand, the Secretary-Treasurer of the Roldskov Trolls, had kept his promise.

"And that, my dear Tomas Graa, is how I arrived in Copenhagen wrapped up in the Danish national colours. And that is how I became the Kongekat. At least, that is the short version of my story. If you're interested, some day I'll give you the long version. I've omitted many behind-the-scenes details that include two minor scandals of international dimensions."

15

Lunch at Villa Fredenskjold

Thomas and Lucas traveled to Nykoebing for their appointment with Helger Helgerssen, the collector of rare books.

Nykoebing S (the S is for Sjælland) is in the Odsherred region, a peninsula that juts into the cold waters of the Kattegat. The Odsherred is full of sand, beaches, summer homes, camping, sailing, and amusement parks. On the north shore of the peninsula, near the village of Havnebyen, Helger Helgerssen's grandfather built a villa in the 1890s. Its extensive front lawn slopes down gently to the sea, and one could think that on a clear day one might see all the way to Sweden.

Thomas and Lucas were picked up at the Nykoebing station and driven to Villa Fredenskjold by Mr. Helgerssen's secretary.

Mr. Helgerssen greeted them at the door. He was a tall man, gaunt, and some years older than Lucas. As soon as he spoke, Lucas recognized from his English accent that Helgerssen was a fellow Cantabrigian. This created a certain mutuality that would otherwise have been impossible.

"Peterhouse," said Mr. Helgerssen, and gave the dates when he'd been there.

"Emmanuel," countered Lucas, "as an undergraduate, but I've been a Fellow at Pembroke for some while now. And Thomas Gray has been at Pembroke almost as long as I."

"Yes, when my secretary told me about your call, I recognized your names. The ibn Kurra story made quite a splash here. Your pictures were in the papers."

Thomas pretended not to hear. It was very hard, though, for her to ignore praise.

Mr. Helgerssen showed them over the villa. They went into a billiards room where a pair of oars had been mounted on the wall.

"You rowed for Peterhouse?" Lucas inquired.

"Absolutely." They went into the dining room for lunch. "Mrs. Helgerssen is sorry she can't be here to meet you. She remains in Milan. Now tell me," he said rubbing his hands, "what can we get for the pussy cat?"

Thomas Gray snarled.

"Don't trouble. Just a bowl, perhaps. I carry a box of Whiskas in her bag."

At lunch, the conversation between the two men got down to business.

"My great-grandfather knew Andersen, of course. There are some family traditions about him. An odd sort. Very. Hobnobbed with kings and millionaires but could never forget his father was a shoemaker. Talked about the nobility of talent. That sort of thing.

"Some people got the idea that a mere shoemaker's son would never have been taken on by the wealthy as a protégé the way he was taken on. He must therefore have been something else: the bastard son of the King, say. How could a shoemaker's son have been such a fine literary stylist? That sort of argument. The somebody-else-wrote-Shakespeare sort of argument. But nothing in it. Nothing."

Lucas Fysst broke in: "I take it that in Denmark, Andersen is regarded as more than just a writer of children's fairy tales?"

"Oh yes. Much more than that. A great stylist. I myself never cared for him. Never tried to collect him. Actually, I doubt if there's anything left to collect. It's all in libraries and museums. Of course, the recent theft of Anderseniana from the Odense Museum opens things up a bit.

"Incidentally, would you like to see my collection? I'm sure you would. The main part is in Copenhagen, I'm afraid. A few odd things are here. Some Aldine Press issues. Some Tycho Brahe. But mostly books about Maria Stuart. Mary, Queen of Scots. My father started it. I've been able to add more from time to time."

"Seems an odd topic, Mary Stuart," Lucas commented.

"All collecting is odd, when you come right down to it. Collecting becomes a mania. It's often a competition between collectors or between collectors and museums whose directors have the collecting bug. Drives prices sky high. That's part of the game. One has to have a few loose pounds, you know."

"But why Mary Stuart?"

"Well, her husband, the Earl of Bothwell, is buried in Faarevejle, a few miles from here. My father, who knew the minister of the church there, got interested in the story. Look, why don't I show you a few items and then we'll take a drive and pay our respects to the Earl."

They walked into the library and to a glassed-in bookcase full of lovely bindings. Mr. Helgerssen took out a key and opened the case.

"Here, let me show you a recent acquisition of mine. James Plumptre, Cambridge, 1796. Title: 'Observations on Hamlet'. Condition: Quite good. A little foxing—discolouration on the lower edge. Puts forward the idea that Shakespeare wrote Hamlet as an indirect accusation of Mary. For Queen Elizabeth's eyes, you know. Brings in the Danish connection in a different way. Rather amusing, what?"

. . .

Helger Helgerssen drove Lucas and Thomas to the Castle of Dragsholm at Nekseloe Bay on the west coast of the Odsherred. "The castle," explained Mr. Helgerssen, "is the oldest secular building standing in Denmark. It's not a castle in the sense of round towers and crenellations. If its stones were modern, one might think Dragsholm to be of modern design.

Lucas knew his Scottish history and reviewed it in his mind.

In a dungeon in Dragsholm, the Earl of Bothwell, third husband of Mary Stuart, Queen of Scots, had been imprisoned. Chained for periods of time, he went mad and died. Frederik II of Denmark had agreeably incarcerated him, hoping to play the "Bothwell card" and recover the Orkney Islands from Scotland. This did not occur.

Lucas Fysst shivered as he recalled Bothwell's story. If the shoe had been on the other foot, he thought, Bothwell would have done as much to his enemies. In the days when earldoms and kingdoms were private inheritances, when differences of religious doctrine often camouflaged other animosities, it was kill or be killed on the greasy pole of possession.

The Castle of Dragsholm is now a small hotel with facilities for small conferences. What irony! Lucas thought; his conference might have booked into Dragsholm. At coffee break time, he mused with a shudder, the conferees might have gone down to the dungeon to smoke, eat a wienerbroed, and discuss briefly the whereabouts of Bothwell's soul.

And the body? Helger Helgerssen informed Lucas the body of Jamie Hepburn, Earl of Bothwell, lay in a crypt in the Faarevejle Church, a few miles to the north. Disinterred some time in the late 1800s, it was found to be remarkably preserved. For a number of years, Bothwell was on display in an open casket, like an open sandwich in the window of a Copenhagen frokost restaurant. Picture postcards are available throughout the Odsherred. Now he has been closed up and lies in a plain, mounted box paid for by the Scottish Historical Society.

If Bothwell had prevailed in his native land, thought Lucas, what a magnificent tomb and marble effigy in the Italian style would have been provided on his death! What elaborate laudations in Latin would have been carved into the side! What falsehoods would have been proclaimed, for in lapidary inscriptions, as Dr. Samuel Johnson said with his tongue in his cheek, "A man is not on oath."

Yes. The coffin was of modern design. There was even a guest book on a nearby shelf. Who had visited Jamie Hepburn recently? Not too many people: Inger Greta Drewsen of Nykoebing and Jens Peter Hermansen of Holbaek. A comedown for one whose life was led among kings and queens and the noble earls of Europe. Lucas signed the book," L. Fysst, Waterfen St. Willow, U.K.," and pushed a Danish crown into a collection slot. While his back was turned, Thomas Gray leaped up out of her shopping bag and sat on Bothwell's box. A cat should sit on a king's coffin from time to time to make sure that the restless, historic spirit within does not mingle with the life without.

Helger Helgerssen drove his guests to the railway station in Holbaek.

"Good-bye. Good luck to you. Do visit me in Copenhagen. Sorry not to know anything helpful. Well, one thing occurs to me now. There is one dealer you ought definitely to contact."

"Who is that?"

"Dr. M. A. Kalamos in Vienna. I've gotten a number of things through him. Very helpful man."

We have now come full circle, thought Thomas Gray, and she went to sleep in the shopping bag.

16

Diamonds Around Her Neck

At 23:12 exactly, Thomas Gray sat on the window seat in Nyhavn 18 and watched the pedestrians below. Swedes carrying tote bags loaded with wine walked in groups toward the last ferry of the evening to Malmoe. When Thomas finally turned around, the ghost of Hans Christian Andersen was sitting in the corner next to her. His back was to the desk. He had a scissors in his hand and an enormous piece of paper in his lap. He was busy making the kind of paper cutting children make when they fold a sheet many times and make incisions at random.

Thomas watched the process. First, H.C. folded the sheet over several times. Then he made cuts on the folded sheet. Quite miraculously, when the paper was unfolded, it exhibited no symmetry whatsoever.

Now, that's a real trick, she thought.

Then he cut away a complete interior section. Instead of separating and falling to the floor, it reassembled itself on the edge of the paper. As the work progressed, a duck appeared, then a pierrot, two ballet dancers, a troll, and an angel.

H.C. (looking up from his work): How are you getting along with your research, Fru Graa?

T.G.: Is it you, H.C.? And if it is you, greetings.

H.C.: Thank you for your invitation, Fru Graa. We spirits don't often get invitations. When the Elf King gave a grand banquet for the Norwegian trolls, he invited all the Danish trolls and goblins, the merfolk, and the various churchyard irregulars, who are all disgusting creatures, really. But he did not invite the ghosts. So there you are.

T.G.: Agreed. I've heard about that. A bad case of discrimination. And retribution undoubtedly set in rapidly.

H.C.: And how is Himself?

T.G.: Himself is wallowing in bafflement. Now let's get down to business. When will you show your hand? Where is your manuscript? Lost in Copenhagen? Our search for it is still on.

H.C.: Ah cat, it's for you to say where it is. I'm beyond direct knowledge or action.

T.G.: Well, I've nothing of any substance to report, old spook: searching, scouring, scratching, poking about. Mostly leg and claw work, you know. We've been clever. We've been sly. We've enlisted help. We've been duped. But we've found no manuscript as yet.

H.C.: Perhaps you need less leg work and more brain work.

T.G.: Spoken in the true continental manner. Theorize first, poke around after. But isn't that out of character for you?

H.C.: How would you know where to poke if you haven't formulated a theory of where and in what manner to poke? Spin tales, make ducks talk all you like if you want, but . . .

T.G.: Leave the deep philosophy for us professionals, H.C. How about giving us a break? After all, it's your reputation at stake. Do you want to be perceived as an anti-cat bigot; *in saeculum saeculorum*, as my coworker, Fysst himself, would say?

H.C.: Sorry, grandma. I am powerless to act against the ground rules that govern transmission of transcendental knowledge.

T.G.: Oh, grandma, is it? How do you figure that?

H.C.: Come now, grandma. And in the fourth generation, too. How many in the litter you had before you took the pledge of celibacy, poverty and scholarship? Four? Five? And the four or five—I know you've lost track—have each had at least four or five, possibly many more—and those, in turn, say four or five. Why, you're now a veritable multiplicity, grandma.

T.G.: I didn't hear you clearly. Speak up a bit. Considerable felicity, you said? That's the conventional wisdom. But don't change the subject, H.C. Now how about it? Throw us a hint. Where's the manuscript?

H.C.: Look in the Gule Sider, the Yellow Pages.

T.G.: Jokes.

H.C.: You may ask me until the last trumpet blows, if you like. The knowledge simply isn't vouchsafed to me.

T.G.: Let's do a little deep memory therapy. There's a picture of you just over Dr. Fysst's desk. You are old. You are looking out the window. This window, even. Thinking. Reminiscing? What were your thoughts at the time? Can you say?

H.C. (releasing the paper cutting which flies upward and dissolves in the air): Yes, cat, I can say. I was wondering who I really was. I was thinking that once I was poor, and through my tales God raised me from the dunghill to sit among the princes of the world. And sit I did among the princes. And it was heady food for the son of a shoemaker and a washerwoman. I was a sensitive swamp plant that had blossomed.

My father? He was an atheist. Do you realize what it meant in those days to be an atheist in a small backwater village? When my mother heard his declaration, she took me to the woodshed and threw a cloth over me to protect me from the malign spirits that must have been hovering about to put such words of disbelief into my father's mouth. When I was eleven, my father was buried in a pauper's grave.

But what was I, really, other than an ugly, awkward, ungainly, tall, vain, cranky, morose teller of tales? A hypochondriac! An occasional and not very successful novelist and poet! A failed playwright!

And I came to love women, but they could not love me. There was Riborg Voigt, and she married someone else. How could she marry someone who'd been born in a hovel? "Goodbye. Goodbye," she wrote, "I hope you will be happy as before." I kept her letter on my person all my life.

And then there was Jenny Lind, the famous soprano. I pursued her all over Europe, even as I was dining with kings. But she wouldn't have me.

The world laid flowers at Jenny's feet and kings placed diamonds around her neck. But she wouldn't have me. I sent her my portrait. I sent her my poems. Perhaps I should have pressed my suit more ardently than I did.

Two years later, she came back to Copenhagen and did a performance at the Royal Theater, just around the corner from here. Well, it's a new building now, of course. While she was in Copenhagen I was by her side constantly.

And when she left for her engagement in Stockholm, she gave a farewell party at the Hotel Royal, a few streets away from here. (Yes, this part of the city is full of ghosts for your present ghost.) Bournonville, the great ballet master, rose and proposed a toast.

"All the men in Denmark want to be her brother," was what he said. And then Jenny responded, "Oh, that would be too many. Andersen, will you be my brother?" And I knew then that it was over. I would have pursued her to Stockholm, but the battle was lost in Copenhagen.

I recovered, of course. And I put Jenny Lind into a story. One of my most popular, as it turned out. I'm happy for that. She is my nightingale.

All my animals, you see, are really people. And I'm also one of the animals. Often.

Yes, Fru Graa, you're interested in cats. Very appropriate. But whom could I have cast as a cat? All the cats I ever knew were ordinary. Ordinary. Not like yourself; not clever at all. No coronet sparkles over your head, Fru Graa, but the nobility of intellect shines through.

T.G.: You tell a sad story, H.C. And it has more than a touch of self-pity. "I was so talented," you say, "I was so worthy of being loved. And I was so unappreciated. Look how I struggled and suffered to achieve prominence." Are such feelings necessary now? In your present state?

H.C.: I cannot shake them off.

T.G.: Will you come again? To give me just a wee hint about the lost manuscript?

H.C.: Unnecessary. What would be gained? And yet . . . no hint. No hint. Not possible. Farvel. I must diffuse.

The spirit of Hans Christian Andersen contracted in the space between the desk and the fan-back chair. It then moved to the window seat, centered itself in Thomas Gray, paused for a moment, and disappeared.

17

Tea at Amalienborg

The arrangements set in motion by Mewsli the Kongekat culminated in the arrival at Nyhavn 18 of an envelope addressed to Dr. Fysst Lucas, bearing the Danish Royal Crest.

The letter inside, signed by the Secretary to the Queen, expressed the Queen's sincere hope that Drs. Fysst Lucas and Thomas Gray would be able to come to tea at the Royal Palace on such-and-such an afternoon. R.S.V.P.

The invitation was R.S.V.P.'d in the affirmative.

On the appointed afternoon, Lucas and Thomas walked to the Amalienborg Palace, ten minutes from Nyhavn 18. After their cre-

dentials were presented, they were escorted by the Queen's Secretary and admitted to an informal reception room. There were soft chairs, sofas, a low coffee table, a small bookcase, and a small rug on the floor with the Dannebrog—the Danish national emblem—woven into it. The Secretary introduced them to the Queen and to the Prince Consort.

"Dr. Lucas? I've read about your cat, and when I heard that you and your cat were both here in Copenhagen, I wanted to meet you."

"My coworker, Ma'am. And the name is Fysst, Ma'am, F-y-s-s-t." Lucas used the mode of address proper for English royalty. He reasoned that it would be appropriate in Denmark.

"Yes. Sorry."

THE QUEEN: Are you both comfortably settled in Copenhagen?

L.F.: Quite, Ma'am. Nyhavn 18. Thomas has quarters in Fyrskib Nr. XVII. The Royal Coast Guard, I believe. Splendid setup.

THE QUEEN: Then we are practically neighbors! And have you met many of us Danes?

L.F.: Yes, quite a few by now.

THE PRINCE: How are you making out with the Danish language?

L.F.: Quite well, Sir, with the written language. Miserably with the spoken language. I keep thinking that if I could only fine-tune what I hear, as one fine-tunes a radio, then all would be comprehensible.

THE PRINCE: Yes, there are jokes to that point. And what do you think of the Danish national character? (He looks at his wife and chuckles.)

L.F.: Quiet. Modest. Sometimes to the point of self-deprecation. Occasionally, that's a camouflage for the opposite of modesty. Perhaps the thought is common that if one shines too brightly, one invites jealousy.

THE PRINCE: It's been said that Danes think rather little of their own work until the outside world tells them it's of first quality.

THE QUEEN (laughing, to Thomas Gray): And from your woman's point of view, Fru Graa?

THOMAS GRAY (looking toward Lucas and indicating her answer to him): From my point of view, Majestaet, which is at a lower level, the Danish character can be summed up in one word: sensible. From my point of view I see lots and lots of shoes. Danish women wear sensible shoes. I would venture that Danish women have the happiest feet in Europe.

L.F. (answering for her): Thomas finds Denmark very comfortable, Ma'am. She is working very productively here.

THE PRINCE: Yes, Danes tend to be hyggelige. Informal. Cozy, *Gemuetlich*, but not in the cloying sense.

THE QUEEN: I hear that you are involved in a research project here in Copenhagen. If it's not a breach of confidence, can you tell about it?

L.F.: It would be my pleasure, Ma'am. We are on a search. The reasons that gave rise to the search are a bit tedious, but we are trying to locate a lost and unpublished manuscript of H. C. Andersen. We have reasons for believing that there is such a manuscript and that when we have found it, it will establish Andersen's attitude toward cats as entirely favourable. Perhaps not generous, but acceptable. In his published work, his attitude toward cats is—well, how shall I put it—rather lukewarm.

Published work is what it is. One cannot alter history, although the attempt is made every day of the year. On the other hand, it would be important to rehabilitate what otherwise might be perceived as a blot on a splendid reputation.

THE QUEEN: You are certain that there is such a manuscript?

L.F.: I have a strong intuition that such a manuscript exists.

T.G. (via Lucas): We know there is such a manuscript, with a certainty that exceeds indubitability.

THE PRINCE: Have you talked to Dr. Ahlborg, the Curator of the Andersen Collection?

L.F.: We have. He was most helpful to us. But his collection yielded nothing. I use the word "his" not in the personal sense.

The tea things were wheeled into the reception room by two servants. On the tea wagon were a magnificent silver tea set, Flora Danica porcelain teacups, and a tiered porcelain pastry tray, on which was arranged an assortment of sweets that was absolutely top of the line, konditorially speaking.

L.F. (self-consciously speaking out of turn, for he had read somewhere that one did not speak to royalty until spoken to): Hasselnoeddefloedsnitter! I am transfigured!

THE QUEEN: Our pastry chef is local, but he worked for awhile at Demel's in Vienna. I think we now have the Viennese beaten.

L.F.: You have indeed, Ma'am. Now tea, in my view, Ma'am, is the most glorious, the most civilized, the most relaxed meal of the day, and I'm sorry to report that we English are allowing the tradition to fall into disuse. And tea with hasselnoeddefloedsnitter . . . well, one cannot imagine . . . In years to come, I shall think: Bliss was it in those afternoons to have been alive.

Lucas broke off a piece of snitte and set it on the floor for Thomas Gray.

THE PRINCE: Would Fru Graa like some Whiskas? The Kongekat lives here, you know. We're well stocked.

T.G. (via L.F.): No. We're fine, thank you very much. And I use the plural "we" in the royal sense.

L.F.: Very kind, I'm sure, but no need to fuss about Thomas.

THE QUEEN: I think it would be a wonderful thing to reestablish Andersen's reputation. Is there any way we can help you?

L.F.: If you could institute a thorough search among the manuscripts of the Royal Accessibilia, I should appreciate it greatly.

THE PRINCE: I shall see to it. I shall call in Dr. Ahlborg and he shall do it.

THE QUEEN: Will you have another snitte, Dr. Lucas?

L.F.: I will indeed, Ma'am.

After the tea and pastry were consumed and a few routine politenesses expressed, the Royal Couple rose.

THE QUEEN: My secretary has your number, and if anything is found in Amalienborg, she will call you. In any case, let me hear from you again before you leave Copenhagen.

The two visitors departed from the Royal Presence. Some days passed. Then the Queen's Secretary rang up Nyhavn 18 and reported that Dr. Ahlborg had performed a thorough search and found nothing.

18

Advanced Procedures

Lucas Fysst sat with his back to the desk and looked out the window over the canal. "You must enlarge on the doctrine of supreme I-consciousness, Swami." His mind was not really on the question he just asked.

"You mean parayogabhinivistasya, I assume."

"Yes. That is the doctrine I mean."

"It says in a commentary on the 45th Sutra of the *Third Book of the Siva Sutras* that a yogi who is absorbed in supreme I- consciousness has both an inward and an outward awareness of the divine. The yogi has an experience where he is inwardly absorbed in the supreme divine consciousness and experiences the world as identical to his own divine consciousness."

"Yes, I see," Lucas Fysst mused. "Now, I don't mean to pass from the divine to the ridiculous, but, Swami, is it within the scope of your professional activity to act as a diviner? I am looking for a lost manuscript, Swami, and I thought perhaps you could help me locate it."

"No. It is not within the scope of my knowledge to find lost objects. That kind of activity lies in the practical world and is within the scope of a competent astrologer."

"An astrologer?"

"The task is routine for interrogational astrology. Now I can supply you with the name of a competent, adequate, if not brilliant, local astrologer here in Copenhagen. I've already consulted him on a personal matter. His name is Mr. Gopinath Singh, and he practices at Krokodillegade 28B."

Lucas went to his desk and opened his Krak atlas of Copenhagen. "Krokodillegade. Yes. Let's see . . . Krogaardsvej, Krohaven, Krohns Vej. Here it is. Krokodillegade. Page 148, G1. Hmm. Near the Citadel. A ten-minute walk from here! Why, it seems everything in Copenhagen is a ten-minute walk from Nyhavn 18!"

"Do not think it outrageous, my son, to learn that the navel of the universe is in all places simultaneously. Now let us return to the knowledge of the supreme I-consciousness."

"My thoughts have wandered, Swami, but they now return."

After the Swami had left, Lucas found Mr. Singh's number in the phone book and rang him up. The call was taken by a young woman with a faintly Eastern accent who booked him for 14:30 on the following Friday. She apologized profusely for Mr. Singh's very tight schedule.

Friday afternoon came, and Lucas walked down the Great Royal Road to Crocodile Street. Krokodillegade 28B was to the rear in a long complex of row houses.

Admitted to a reception room by a secretary, Lucas was surprised to find it decorated in the modern Danish style: no dark curtain of oriental beads to pass through, no zodiacal signs attached to the walls. In one corner there was a personal computer and printer.

Mr. Gopinath Singh appeared and the two men shook hands.

"Ah. Reverend, I see. I am glad to be of some service. I have helped many men of cloth."

"I'm sure you have. Incidentally, you have a very attractive office here."

"Thank you. We are quite up to date. Planetary positions, oppositions, conjunctions, ascendencies, et cetera, are all computerized now. There are now quite reliable software packages in the astrological line. They free us for the interpretive portion of our work, which is the far more creative and challenging portion."

"I should think they would be. I've heard the claim made quite generally for computers."

"Now, Mr. Fysst, can you tell me the exact year, month, day, hour and minute of your birth? As exact as possible. I realize one doesn't know these things too well in the Western world."

Lucas Fysst obliged him with the information. As Mr. Singh had surmised, Lucas did not know the hour or the minute.

"Now, Mr. Fysst, tell me the nature of your inquiry."

"I am looking for a lost manuscript."

"When did you lose it? As exact as possible, please. Tell me the year, month, day, hour, and minute of the loss."

"I didn't lose it. It was lost by Hans Christian Andersen. He's a very famous man in Copenhagen."

"Yes, I know the name quite well. I believe he is a movie director. I saw one of his movies when I was a child. With Danny Kaye. Now Mr. Fysst, we are confronted with a fundamental difficulty here. The normal procedure of interrogational astrology would be to have the person who suffered the loss initiate the inquiry. I must see and talk with such an applicant. I must obtain the vital data. In the profession, we call these the initial conditions of the event under investigation. Shall my secretary make an appointment for Mr. Andersen?"

"That's not possible."

"That's up to you, of course. In such a case, working by proxy would involve a lengthier procedure and involve more uncertainties. The fees would be doubled. Approximately when did Mr. Andersen lose the manuscript?"

"If he lost it, probably sometime between 1845 and 1875."

"How's that?"

"You see, H. C. Andersen died in 1875. I'm inquiring on his behalf only in a larger sense. Now it's quite conceivable that at his death, the manuscript was simply among his papers and was somehow lost to the world at a later date."

"What you ask for is very unusual. But not absolutely unheard of. The procedures now become exceedingly complex. The fee will be still higher."

"Roughly how much?"

"Say 1000 Danish kroner. And not covered by the National Health Service."

"That is reasonable. You may proceed with the inquiry."

"I shall first need to interview the ghost of the applicant."

"If you say so. I assume you have the means for bringing about such an interview."

"My dear Mr. Fysst, that would be *your* responsibility. I'm a man of science, an astrologer, not a necromancer. Farvel."

"I'll do what I can. Farvel."

"Farvel again, and may your arrangements be successful."

Back in Nyhavn 18, Lucas Fysst addressed Thomas Gray. "You may wonder, old thing, whether in attempting to set up an interview between a ghost and an astrologer, I am acting in a heterodox, unscientific, irresponsible, and completely bonkers way. As far as the first is concerned, the Western Church, over the past two thousand years, has had mixed feelings about the existence of ghosts and about astrological procedures. At the moment, it is rather down on both.

"As to the second, orthodox science denies the existence of ghosts and declares astrology to be a failure at best and a sham and a delusion at worst. But orthodox science fails to consider the element of human psychology. The mind is an area where orthodox science has thus far had a miserable record.

"The human heart wants to know many things and asks many questions. What path should I take in life? Whom should I marry? When should I undertake a journey? Where can I find a lost manuscript? Will I become famous? Rich? How long will I live? To expunge such questions from consciousness would be less than human. To expect a physicist or a social scientist to answer such questions would be absurd. What is at stake here is the self—the individual self, not a social average.

"The astrologer is prepared to answer all these questions. The astrologer works partly within a tradition and partly by psychological intuition. That's why Mr. Singh asked for a direct interview.

"So we may have recourse to whatever comfort we can find within whatever systems are available. And I turn to unconventional procedures in the hope that, however ill-based they may be, they will stir up in me—or in you, dear Thomas—additional insights that will allow for a breakthrough."

Thomas Gray listened to these words with considerable skepticism. But she gave in, and that night, sitting in Nyhavn 18 on the spot where Andersen's desk had been, she got in touch with his ghost and did her best to have it consent to an interview with Mr. Gopinath Singh.

The apparition arrived shortly after midnight. A heavily laden mahogany bookcase materialized along with framed photographs on the wall. A round table emerged from nowhere, and on it was a sculpted head of Jenny Lind. Hans Christian Andersen was in his usual black frock coat and vest, baggy trousers and stovepipe hat. He carried a small oil lamp whose wick was glowing. He set the lamp on the table and his hat on a chair. The quartz clock on the wall stopped ticking.

> H.C. (his voice sounding particularly hollow): Aah, it's you again who put through the call.
>
> T.G.: Quite. Your metasouled friend, as you once put it coarsely.
>
> H.C.: Well, Fru Graa, what shall be the subject of our conversation this time?
>
> T.G.: Time? What meaning can time have for you? I assume you're using the word "time" in its infradental sense.
>
> H.C.: Infradental? A medical term? You have suffered, as I have, from the toothache, Fru Graa? Oral surgery, perhaps? A root canal on the lower jaw?
>
> T.G.: I will explain, though this is probably carrying coals to Newcastle. Reality is split into two hemispheres: the tran-

scendental hemisphere and the infradental hemisphere. Your world belongs to the transcendental, the world of my collaborator, Lucas Fysst, to the infradental.

H.C.: And where is your world? Where, exactly, do you fit in this classification?

T.G.: My dear sir, I am a cat. I am absolutely unclassifiable in any taxonomy!

H.C.: I will take that as a personal triumph for you. However, I assume you rang up for business and not to haggle about classification schemes.

T.G.: Yes. My collaborator, Lucas Fysst, has consulted a certain Gopinath Singh, who practices interrogational astrology.

H.C.: I have heard of this gentleman. And to speak plainly, I believe that omniscience is his foible.

T.G.: In the case of a lost object, to wit and viz., your manuscript, H.C., this Mr. Singh, in order to practice his art, and to recover what was lost, requires certain parameters pertaining to the circumstances of the loss. When the loss was first noticed et cetera.

H.C.: I am familiar with the procedure, and I must tell you that if Mr. Singh possessed a higher level of skill in plying his art, a personal interview would be totally unnecessary.

T.G.: Beggars can't be choosers. Will you allow yourself to be questioned by this Mr. Singh?

H.C.: I will not.

T.G.: Then will you provide me with the parameters for transmittal to Mr. Singh?

H.C.: Parameters be damned. I will not.

T.G.: Your memory has been wiped, then?

The apparition took a large handkerchief from its pocket and mopped its brow.

> H.C.: On the contrary, my memory burns. At the time I wrote the story, I was in heaven and I was in hell. I was both a grand success and the greatest of failures. I sat next to kings and queens; mind you, I, the son of a washerwoman sat next to royalty and the crème de la crème of the world's writers. I often did not know who I was. And then I was ill. With the toothache and with much more besides. Invitations came to me from across the seas. I would be feted and honored. I believed I was going mad. I said no to the invitations and wrote a story to relieve my feelings.
> Most of what I have written is a reflection of myself. The stories lay in my thoughts, requiring only a flowing stream, a ray of sunshine, and a drop of wormwood for them to burst into bloom. Every character was taken from life. I knew and have known them all. In the wienerbroed story, I was the wienerbroed.
> Parameters? Do what you will with what I've just told you. Enough. Enough. Enough.

The phantasm turned down the wick of the oil lamp. The smell of acrid smoke filled the room. Thomas Gray sneezed.

> H.C.: Farvel. Farvel.

The image of Jenny Lind faded, and the clock on the wall resumed its ticking.

19

More Pressure Is Applied

It was a bit of a walk for Thomas Gray to get to Christiania. She paced herself. She left Fyrskib Nr. XVII well after rush hour so that the automobile traffic she had to cross would be less fierce. When she arrived in Christiania, it was well into the evening. She made inquiries with the local feline authorities and found the Swami sitting on the ground behind a kiosk that sold potted plants. She joined him, and together they sat in silence.

Christiania is located on the other side of the harbour passage from Nyhavn. It is a walled enclave, formerly a military barracks, inhabited now by perhaps a thousand students and squatters who, in different degrees and manners, practice an alternate lifestyle. One hears stories that the government of Denmark regards Christiania as a social experiment and that its laws are not operative within the walls. The place gives the initial impression of a run down carnival, minus the rides. Christiania's reputation for heterodoxy attracts more tourists than locals. The conservative socialist burghers of Copenhagen stay away and would probably be shocked by the high level of graffiti and the scores of dogs that run around loose in Christiania's open spaces.

The bells of the nearby Vor Frelsers Kirke, that wonderful baroque church with a spiral staircase appliqued to the exterior of its steeple, bonged nine times, indicating 21:00. Some strokes sounded as though they were coming from near, others from far. Evidently, the horological machinery had been installed long before the European destruction of A.M. and P.M.

Thomas Gray still sat opposite the Swami. Both were in rapt contemplation. Was communication possible between them: the Swami, whose soul, he hoped, was approaching the ultimate absolute, and Thomas Gray, in whom resided the soul of Vyaghrapada, the Swami's late guru, according to the expressed view of the Swami? The Swami began a rhythmic recitation of O Maa, O Maa, Om, Om, which he eventually discontinued. He then spoke to Thomas Gray.

"I honor you, and I shall ply you with tidbits, O Guru-Cat, but please go away. Return to where you stay. Return to the Reverend Fysst, who needs you. Do not remain with me and make requests that extend beyond the reach of my regimen of contemplation."

Thomas Gray remained silent.

"Why do you not depart in peace, O soul of my soul? Why do you reproach me for my inabilities, for my myriad imperfections?

The illumination I have received is as black as the night when measured against the unspoken demands flashed to me by your green, penetrating eyes."

Thomas Gray's whiskers twitched in ticlike fashion.

"I have given you my best advice: Inquire from Mr. Gopinath Singh. And if, as you indicate to me, these inquiries resulted in emptiness and frustration, take it as a sign from essential nature that she is unwilling to deliver up her secrets. Go, O Cat. Do not try to upset the order of things. Direct your steps out of these confines. What wouldn't I endure to be able to give you what your heart desires? Now what is the lesson? The lesson is that we should reduce our desires to the vanishing point."

Thomas Gray rose and plotted an exit path to the exterior of Christiania that avoided as many of the random dogs as possible.

Well, I tried, she thought to herself, as she made her way down Prinsessegade toward the main thoroughfare of Torvegade. A number 31 bus sped past. A full moon shone over the dense housing of Christianshavn and cast dark shadows.

It often seems, said Thomas to herself, that walls are built to create chiaroscuro. They give illumination its strength, even though such illumination is always imperfect.

The Search was back to square one.

20

"Perhaps YOU Have!"

At 15:03 one afternoon several weeks later, the telephone in Nyhavn 18 rang.

"33 91 22 18; Fysst here," said Lucas, in his best crisp British manner.

A woman's voice came from the other end. The voice started to speak in Danish, then switched to English. "Have you seen the ghost of Hans Christian Andersen?"

Lucas Fysst thought to himself, A crank call. A hoax. I shall hang up. No, he then said to himself, I'll not hang up. Nothing ventured, nothing gained. I shall answer rather ambiguously.

"It depends on what interpretation you give to the word 'see,'" he answered in his best philosophic manner.

The woman knew English, but not sufficiently well to understand the word "interpretation," and Fysst's reply baffled her.

"Who are you?" she asked.

Who am I indeed, young lady, Lucas thought to himself. Well, she may not be young, but she sounds young. Perhaps it's Stephanie, the young girl I was with on the train from Hamburg. Perhaps she's pulling my leg. No. I think it can't be Stephanie.

"It's you who initiated the call. You tell me who you are," Lucas asked.

Without hesitation, the voice answered, "I'm Karen."

"Froeken Karen, I'm Lucas Fysst."

With more than usual curiosity, Thomas Gray hopped onto the telephone table and listened to the conversation.

"Well, Karen, why did you ask me whether I'd seen a ghost?" Lucas asked.

"I work for Danish National Television in Soeborg and . . ."

As soon as Thomas Gray heard the word "television," she became agitated and nearly fell from the table. Lucas also became excited and, not waiting for the woman to finish her statement, launched into a long and garbled story. He concluded the garble by saying, "If you could visit me in this flat (you know the address because you called me and not the other way around), I'll explain my relations with Andersen. Everything."

"Well . . . let me call you back. I'll call back in . . . oh, about ten minutes."

At the Danish National Television headquarters in Soeborg, Building A, Karen consulted the Head of Studio, Lars Jakob Pedersen.

She told him of her idea to put together a little show dealing with a number of reported ghost sightings.

"Such as?" asked Pedersen.

"First, the ghost of H. C. Andersen, sighted just outside Nyhavn 18, where he lived a hundred years ago. Second, when the remains of Bothwell were transferred to a new coffin a couple of years ago, the locals in Faarevejle reported emanations and rumblings."

"Who is Bothwell?"

"The Earl of Bothwell. Queen Maria Stuart's husband.

"Any other cases?"

"There's always the ghost of Hamlet's father at Helsingoer. Constant sightings."

"How do you hear about these things?"

"People call in to TV 1 or TV 2 and tell us."

"So what's your problem?"

"I got the telephone number at Nyhavn 18. I rang up. A Britisher answered. I asked him whether he'd seen the ghost of Andersen. He gave me a long story about a cat, a lost fairy tale, a swami, and an astrologer. He told me that he and his cat had had tea with the Queen. I think he's a nut. He said if I came to his flat he'd show me some things and tell me a lot more. Shall I take a chance?"

"Hvo intet vover intet vinder." Nothing ventured, nothing gained.

"O.K. I'll check back here after a half hour. If you don't hear from me, send the rescue squad."

. . .

The phone at Nyhavn 18 rang again. Karen said she would be there at 18:00.

"Splendid. I will tell you all," answered Lucas. "No. Some. All is much too complex. All would take too long. And I must go to the opera tonight. 19:00. *Tosca.* The Opera House is just around the corner."

At 16:40, Lucas Fysst walked around the corner to the Magasin du Nord and bought a white shirt and a snap-on black bow tie. Very appropriate for the opera, he said to himself. I shall go in mufti. His black ecclesiastical shirt and white collar would have been appropriate, but he had something else in mind.

At 17:30, Lucas Fysst went to the kitchen and fixed himself a light snack.

At 17:45, he put on his dark suit, white shirt, and black bow tie. He looked at himself in the mirror. Perfect for Andersen. Would Karen show up? It struck him now that the telephone call might have been a deception. A hoax.

At 18:00 the bell to his flat rang. He pushed the buzzer to release the downstairs lock. He heard the click of heels on the steps.

"I'm Karen."

"Hello. I'm Lucas Fysst. Please come in and let me show you my private little Andersen museum."

Karen appeared young and alert. It was raining out. Lucas took her rain coat and put it in the closet.

By 18:10, he had shown and described to her all the photographs and artwork on the walls. He told her to notice, particularly, the picture of H.C.A. by the window.

Thomas Gray entered the room quietly.

"This is Thomas Gray. She is an excellent investigator in the manuscript line. Together we have discovered a new Andersen story. Absolutely unpublished."

Karen was confused. "Together?"

"I haven't told the Queen yet."

"The Queen? You haven't? Where did you discover whatever it is you discovered?"

"In the Kongelige Bibliotek, naturally. It's been there since 1897. But not in the principal Andersen Collection."

"It has?"

"Would you like to hear the story we've found?"

Karen thought, Should I excuse myself now and ask to make a phone call to Pedersen to come and get me out of this? She hesitated.

"Of course you would," said Lucas Fysst. "Sit down at that little table and I'll sit opposite you."

Thomas Gray watched from beside the low table. Lucas opened his notebook and read aloud, slowly. Karen listened, absorbed.

By 18:20, the reading was completed. Lucas put his notebook back on the desk.

Karen did not speak for some while. Outside, the wind and rain increased in intensity; the drops beat on the window.

"You've found a story. And did you see the ghost of Hans Christian Andersen?"

Thomas Gray's whiskers twitched violently. The sky darkened. Lightning cracked down on a nearby roof.

Lucas Fysst stared back at Karen and she, in turn, looked deep into his eyes. Lucas answered, "Perhaps YOU have! Perhaps you have seen his ghost!"

For a moment, Karen was unnerved. She rose. She thought she ought to phone Pedersen. Then she realized, but only partially, what was going on. She regained her composure. She smiled.

"Could you repeat this whole performance?" she asked.

"What performance?"

"Coming to the door. Showing me the pictures. Reading the story. Wearing a tie like Andersen wore. Scaring me out of my wits. It was a performance, wasn't it?"

"In principle, I'd be very glad to, but I've got to get to *Tosca* in a few minutes."

"I don't mean now. In a couple of weeks. I'll come with a camera crew. We'll shoot it all."

"Splendid! But you must feature Thomas. She is very cinematogenic."

"Absolutely. Farvel. Jeg haaber vi ses igen."

"Farvel. Vi ses us igen and again and again."

. . .

At 19:01, the opera curtain rose. Lucas Fysst sat in the center of the parquet and prepared to listen to the sorrows of Tosca.

After the opera, while enjoying a midnight posset back at the Nyhavn flat, Lucas reflected on the events of the afternoon. Thomas Gray, he thought, is a splendid cat. How agitated she was at the prospect of a television connection! Kierkegaard has described Thomas to a T. "The sins of the talented," said that great Dane, "are great and twofold. The talented think they have merited the smartness they were born with. And the talented overvalue by far their accomplishments." Thomas has many virtues. Modesty is not among them.

21

Den Lille Havfrue

Early on a rare evening when the sky was clear, Lucas Fysst said to Thomas Gray, the time has come, old thing, for us to pay our respects to the Little Mermaid. I read in a tourist publication that she is the soul of Copenhagen. I'll reserve judgement on that point. In any case, it's certain that she's the logo of the city."

Thomas indicated willingness.

"Poor girl. Why can't they leave her alone? A few years back her head was sawed off. A strange derangement in the perpetrator. Now, I've read in the papers that the manuscript of her story, her literary birth certificate so to speak, has been stolen. It seems that life for icons is not all beer and skittles."

Setting out from Nyhavn 18, they walked north, past the Admiral Hotel and the Amalienborg Palace. They continued past an enormous fountain with huge sculptures of oxen, fish, and goddesses. Ultimately, on the harbour front at the edge of a park, they came to the Little Mermaid sitting cool and damp on her mossy rock. She was quite close to solid land; one could get to her by stepping over a few stones. Behind the Little Mermaid was the open harbour. Tourists were now crowding back into the last bus of the evening.

Thomas Gray negotiated the slippery rocks and sat next to the metallic maiden.

"I have no camera," Lucas said, "so you needn't pose. Do you know her story, Thomas? No, why should you know it? Listen, and I'll tell it to you. The short version, of course.

"Once upon a time, there was a mermaid princess who lived in the kingdom under the sea. Now when mermaids reach the age of fifteen, they are allowed to rise to the surface of the water and sit on the rocks in the moonlight. Our little mermaid turned fifteen in the winter and sat on the top of a large iceberg, with her loose hair streaming in the wind.

"Somewhat later, she rose up and saw a ship with a young prince on board. The little mermaid fell in love with the prince. A severe storm arose, and all aboard were cast into the waters. The prince became unconscious after his strength gave out, and the little mermaid rescued him and took him to land. Over the following months she saw him in his palace by the sea, and she longed for him.

"Now I must tell you several things about the mer people. They live for three hundred years. On the other hand, they have no souls, and when they die they simply become foam on the water. Humans, on the contrary, have immortal souls that rise up to the stars and live forever. Now if a human loved a mermaid with absolute love, his soul would infuse hers and she would gain a soul, but he would not lose his. So the little mermaid longed for the prince to love her absolutely so that she would gain a soul.

"Of course, there was an obvious impediment: her fish's tail. Humans do not care for such things, nor are tails particularly convenient for life on land. To make a comparison, Thomas, a lady's bustle of the late nineteenth century would be a serious impediment to driving a car. So the little mermaid went to visit the Sea Witch. The witch lived in a frightening part of the sea, full of frogs and snakes and giant squid. The mermaid made her request to the witch (the witch knew the request beforehand, actually), and the witch agreed. The mermaid's tail would be replaced by two human legs.

"But the witch drove a very hard bargain, and the price she asked for this transformation was horrible. When the little mermaid walked on her new legs, she would be in constant pain, as though she were walking on knives. More than this, her tongue would be cut off and she would lose her beautiful voice. And still more, if she failed to gain the prince's absolute love, the moment the prince married someone else, she would turn into foam.

"Well, nothing ventured, nothing gained, and the little mermaid agreed to this bargain. Now, to make a long story short . . ."

"Yes," said Thomas Gray, "cut the frosting. You're not Andersen, you know. Get down to the cake."

" . . . the Prince loved another better than he loved the little mermaid. A date was set for his marriage. The four sisters of the little mermaid pleaded with the Sea Witch for respite. For the price of their lovely hair, they obtained a magic knife.

"'Plunge this knife into the prince's heart,' the sisters pleaded with the little mermaid. 'When his warm blood spills over your feet, they will become a tail again and all will be as before.'

"But the little mermaid, looking down on the soon-to-be-married prince, was not able to do it. She cast the knife into the sea. Immediately, her body dissolved into foam, and the foam evaporated and took on the form of the Daughters of the Air. If, over a period of three hundred years, such beings do a sufficient number of good deeds for the human world, then they may gain a soul."

Lucas Fysst turned around and walked back slowly towards Nyhavn. For a moment, Thomas Gray remained facing the Little Mermaid, soliloquizing.

"O twice-abstracted piece of bronze that purports to be half girl, half fish. O beautiful and yet nonexistent creature of fantasy.

You are so sweet. The world loves you and makes pilgrimage to view you. Are you not a universal icon? Henry Moore, who filled our English landscape and filled your Louisiana Museum gardens with his sculptures, would not have rendered you thus. But that is not the issue; it is, rather: Do you have a soul?

"How can it be denied? One might have thought that having withheld the magic knife from the prince you loved, you would certainly have gained a soul thereby. Not at all. Does the world really know or care to know the story that Andersen, your creator, told? Does the world know that Andersen gave an ambiguous answer?

"Virtue unrewarded is the song that Andersen sings. Over and over again. Is his perverse private character at work in this song? Does his bitterness at imagined neglect reverberate in some small measure in all our breasts? In our regard for this lovely piece of metal, have we rewritten Andersen's story to sedate our own broken hearts?"

22

Bona Fides Established

Karen, the television woman, must have had reservations about the character of the man she had just dealt with. Was Lucas Fysst mad? Was he a genius? Was he, as he pretended to be, a reincarnation of Andersen? Was he, as he claimed, a university lecturer? Or was he simply an imposter? Tea with the Queen? An unlikely story! She rang him up the following day and asked whether she could visit him and his little museum again.

"Why, most certainly."
"May I bring a friend?"
"Why, most certainly."

Karen's friend turned out to be a somewhat older lady. "This," she said by way of introduction, "is my assistant, Birte."

After some trivial politenesses, Karen and Birte went into the H. C. Andersen room and spoke together in Danish. They seemed to be discussing the walls, the windows, the pictures, and the furniture. When they returned to the living room, Karen asked, "What exactly is it you do for a living, Mr. Fysst?"

"Well, I'm a university lecturer, you know. In England."

"What do you lecture about?"

"The history of ancient ideas. Particularly scientific ideas, religious ideas, and their relationship."

"What's that about? I never went to university."

"Oh, it's such a vast subject that I hardly know where to begin. I have an idea. Months ago, Professor Mogens Hvidvin asked me to give the Kierkegaard Memorial Lecture at the University of Copenhagen. I shall be speaking on Thursday afternoon at 14:00, and I invite you to attend. That should give you an idea."

"You don't have to be a student to attend?"

"Oh, no. It will be a public lecture."

"What was the man's name again?"

"Kierkegaard."

"I don't mean him."

Lucas repeated Hvidvin's name and Karen wrote it down in her notebook.

"Can I sit up front?" she asked naively.

"Yes. Most certainly. You can sit up front. You can sit in the back. Anywhere you like."

"I thought that only very smart people sat up front."

"Very smart people sit wherever they like. And so shall you. But I'll say this. Why don't you sit near the door? Then if you don't understand what I say or if you find it boring, why, just get up and leave. It's a common practice at university lectures. It happens every day of the week."

"What will you talk about?"

"My lecture is titled 'Scientific Views on the Immortality of the Soul.'"

"That sounds good to me. Will you talk about ghosts?"

"It's quite possible."

"Then I'll come. Can Birte come?"

"Yes, of course," and looking at Birte, he added, "please do come.

And bring any friend who would like to join you. My lectures rarely crowd the lecture room."

"And can Birte sit with me up front?"

"Why, most certainly," Lucas answered. The two women left Nyhavn 18 looking quite pleased.

On Thursday at 13:45, Lucas Fysst went into the lecture room escorted by Professor Mogens Hvidvin.

"A lady rang me up an hour ago," Hvidvin said. "Said she was from Danish National Television. She seemed to be checking up on your credentials, Lucas." Hvidvin laughed.

"I invited her to come."

"You mean," asked Hvidvin incredulously, "that Danish TV intends to report the Kierkegaard Lecture? That would be a miracle."

"The miracle, like all good miracles, will be tangential rather than direct, if it occurs at all."

Karen and Birte showed up for the lecture. They sat in front. The two professional women stood out clearly against the standardized raggle-taggle dress of the faculty and students.

Lucas opened his lecture with the words: "Odysseus is in the Kingdom of the Dead and meets the ghost of Achilles. Odysseus speaks first (I render freely): 'O happiest man who ever has been or is yet to be, do not sorrow so that you are dead, Achilles. In your day, we honored you with a god's honors, and now I find you are a prince among the dead.'

"And Achilles answered, 'O shining Odysseus, do not make light of death before me. I would rather be on earth as a slave to the poorest man than be king of all the ghosts that are here.'"

The lecture continued for an hour, and the two women stayed well into the question period. Then they came up to the podium, thanked the speaker and left. Lucas never found out whether they had understood a word of what he had said. He suspected that they had sat through his display of scholarship because they were waiting for additional occurrences of the word "ghost," which, alas, did not materialize.

23

Chaos Off Camera

A date and time were set for the shooting in Nyhavn 18. Lucas, speaking to Karen on the phone, said he did not want the discovery of the Andersen manuscript to be part of an evening news report. There were things about it still to be cleared up, he explained. She gave her assurances and said that the shooting would be for a show for young teenagers called *Troldspejlen*. She was the producer and director of *Troldspejlen*.

The crew came at 13:30. Along with Karen, who was in charge, there was a camera operator, an audio technician, and a lighting expert. Birte, Karen's assistant, turned out to be the continuity person. They moved in their equipment. They plugged in heavy cables and extended them over the floor. Thomas Gray thought they lay like large wet spaghetti, and she walked over them quite cautiously.

The crew arranged floodlights and reflectors, and taped translucent sheets over the windows. They rearranged the furniture in the H. C. Andersen room. They removed the clock from the wall, took the phone off the hook, and taped up the doorbell. They jabbered amongst themselves in Danish; to Lucas they spoke in comprehensible, though not elegant, English. Lucas, dressed up in his black suit and black bow tie, was wired for sound.

After a good half hour, they were ready to shoot.

"Will we use a script?" Lucas asked Karen.

"No, we'll just do what we did when I first came here."

It is well known that TV material is shot in tiny bites that are later assembled. Lucas, who had hardly acknowledged the existence of the TV medium, was unaware of this. His idea was that he would lecture to the camera as he might lecture to his students in his course on the historiography of Mesopotamian mathematics.

Karen, his boss for the afternoon, provided Lucas with illumination on the matter, bringing him into the late twentieth century. "Now Lucas," she said, as though she were handling a seven-year-old, "at first the camera and I will be out on the landing. I'll signal you by knocking (the bell being taped over). You open the door and say 'Hello, Karen.' Then say whatever you like. But make it short. And remember, the camera will be on you, not on me."

"O.K. I understand."

Knock, knock, knock. Lucas opened the door. The spotlights on the landing blinded his eyes.

"Hello, Karen. Won't you come in? I should very much like to show you my little Andersen museum."

"Cut. Now Lucas, let's take it again. I don't think you opened the door wide enough. Open it all the way."

Knock, knock, knock. Lucas opened the door wide.

"Hello, Karen. Won't you come in? I'd like to show you my private—well it's not mine really—Andersen museum."

"Cut. Could you smile just a bit when you say 'Hello, Karen'?"

Karen closed the door. Knock, knock, knock. Lucas opened the door and smiled. Perhaps too wide a smile, he thought.

"Hello, Karen," he said. "I'm delighted you could come. I'd like to show you my Andersen museum."

"Good. That worked. Let's get inside your flat now."

"Remember, you must put Thomas Gray on camera," said Lucas.

"Later. I won't forget. No way."

By 14:23, two more bits had been shot. Between each tiny scene, there was hard and elaborate work on the part of the crew: setting up and rearranging. For Lucas it was idle time to do nothing, stay out of the way, and watch.

At 14:36, during a rest period, the Swami came in silently. He ignored the activity and went into the dining room. The two men sat at the dining room table, away from the shooting.

The Swami said, "As long as there are no knowers, nothing can be known. The knower and the known are really identical. It follows, then, that there is nothing which, in itself, is impure."

At 14:44 Karen called, "Mr. Lucas! Come back, please! We're ready."

"Many pardons, Swami. I shall return shortly," promised Lucas, and rejoined the television crew.

Karen ordered him, "Step over the spaghetti. The lights are up now. Start."

Lucas began, "Now here is a photograph of H. C. Andersen in his final years. You see he sits at his desk in the very corner where this modern desk now stands."

Karen stopped him. "Take it again. As you speak, go over to the photograph, and pretend to study it."

Lucas began again. "Now here is a photograph of H. C. Andersen in his final years. You see he sits at his desk in the very corner where this contemporary desk now stands."

At 15:12, Lucas got some more time out. Once again, he joined the Swami. They were studying Sutra 1, Section I, when he heard the sound of heavy steps.

"Howdy, old friend. Pete Williams from Big Bend, Texas. I'm back from Budapest and trying to work up a little ding ding with the born-again capitalists. Good to see you, ol' pal! And this here is . . .?"

"How delightful!" said Lucas, recognizing his old seat-mate from the SAS flight to Hamburg. "You came in time for the shooting!"

"Shooting in Denmark? Waal, now!"

"And you know, I've been thinking about the lesson in Danish I gave you on the plane. Excuse me, Mr. Williams, allow me to introduce the Swami to you."

The Swami made an Eastern gesture of greeting. "Behold! An aspirant? A seeker of the purpose of being?"

"Glad to know you, pal. Now, any friend of a friend is a friend of mine. I don't know where you hail from, pal, but I'll tell you both that I just heard that us Big Bend folks are laying on the biggest-ever chili-off that was ever offed and I want to extend, from Revelda Mae and mahse'f . . . "

"Mr. Lucas! Please. We're set to go," called Karen.

"Yes," replied Lucas. "As I was saying before, Mr. Williams, in the plane, regarding long words in the Danish language, I've found perhaps the longest one known. In a story by Andersen. Yes, Andersen: 'The Shepherdess and the Chimney Sweep.' Put it in your computer and analyze it at your leisure, as they say. Ready?

GEDEBUKKEBENSOVEROGUNDERGENERALKRIGSKOMMANDOERSERGENTEN

A piece of satire on official pomposity, really."

"Mr. Lucas. We're ready. Time to read the new story," called Karen.

Lucas went back to the cameras, leaving Pete Williams with the Swami. Karen set the scene.

"We'll sit at the little round table," she directed. "I'll be opposite you. I'll ask you some general questions about your impressions of Copenhagen, Andersen, Thomas Gray, your discovery, and the Kongelige Bibliotek. We'll get some shots of that separately. Get the cat to sit on the table. The camera will be on you and the cat alternately, not on me. We'll shoot the whole thing twice. Then we'll mix it in the studio. Reach over to the desk. Get your notebook out. Read the story slowly. Close the book. I'll ask you, 'Did you see the ghost?' Remember, answer just as you answered when I first met you. At that point, do something to the cat. Get her to jump off the table."

Thomas Gray heard the instructions. She played her part perfectly, without prompting. In fact, she did it twice. A seasoned actress now, she walked out of the room with complete nonchalance and self-assurance. She went to the dining room, jumped up onto the dining table, and listened to the Swami, who was now providing ancient enlightenment to Pete Williams.

Then came a knock, knock, knock on the door. Pete Williams opened the door and a lady appeared.

"I'm Pete Williams, ma'am. Won't you come in?"

The lady responded in Danish. Williams asked Birte to come and translate.

>LADY: Is the woman from the television station here?
>BIRTE: We're busy. What do you want?
>LADY: I want to speak to the woman. It's a private matter.
>BIRTE: Karen, there's a lady here who wants to talk to you.
>KAREN: Yes? We're very busy.
>LADY: I called you from Sneglerup three weeks ago. I saw the ghost at Dragsholm. Is there a reward?
>KAREN: I'll talk with you later. Sit down in the next room. [Danish for "Buzz off, honey."]

At 15:37 the shooting resumed, this time through the window. They shot the canal, the boats, the restaurants on the opposite side, and the people on the pavements.

>LADY (to no one in particular): Can I be on TV with the ghost? My dog Chokolade would love to see me there.
>THOMAS GRAY: Hssss!
>PETE WILLIAMS: Get Anton Berg's chocolate. They make the best chocolate in Denmark. Denmark's great. But you know what their main trouble is?
>LUCAS FYSST: No. What is it?
>PETE WILLIAMS: Bullet: No pahs, suh. No pahs!

LUCAS FYSST: Pahs?

PETE WILLIAMS: Right! Pahs! No apple pahs! No pumpkin pahs! Whah, in Marie Callender's Restaurant in El Paso, you kin git twenny-four kines of pah!

LADY (in Danish): I was on late duty turning down beds at Dragsholm. I was going down the back staircase to the second floor. The light had blown out.

PETE WILLIAMS (to the Swami): Do you understand Dansk, pal? What's she saying?

LADY: A blue cloud form came toward me from below and said, "Maria, my Queen." I screamed and the cloud went through me. I told the night manager." It's the ghost of Bothwell," he said. [Birte is still translating for the lady.]

SWAMI: It is a profoundly spiritual experience. In the *Vaya Purana* it is written that if an aspirant enters an astral body that appears, the aspirant should deem it a spiritual attainment.

THOMAS GRAY (to herself): Deem. Deem. Deem away.

PETE WILLIAMS: Bullet: Bottom line—Deem to attain spiritual values.

At 16:18, Lucas Fysst, absolutely exhausted, asked himself, How can acting be so exhausting, when it generally consists of saying six words at a time? He answered his own question: By saying it over and over and over again.

LADY (to the Swami): Why don't you wear more clothes? In Sneglerup you would freeze to death. These Copenhageners!
[No response from the Swami.]

LADY: I said, Why don't you put on some warm clothes?
SWAMI: Madame, I do not understand you. I cannot speak Danish. Jeg ikke taler dansk.
LADY: Ikke?
SWAMI: Ikke.
PETE WILLIAMS: Me also ikke.

At 16:28, Lucas said, "Mr. Williams, take this money. Run down to the Illum shop across the Kong—I mean across the square—and buy pastry and more pastry. We must have a party to celebrate and reenergize."

24

The Magic Mirror

It took several days for Lucas Fysst to recover from the fatigue of his maiden acting experience. As he told his wife much later, he had gained a new appreciation of what is involved in that profession. Thomas Gray, on the other hand, was up and around without an ache in her bones. About a week after the shooting, Karen rang up Lucas and said she needed to return to check out a few details. He invited her for supper at the Cafe Italiano across the canal on Tolbodgade.

"So. How long have you been working for the *Troldspejlen* program?" he asked Karen after their orders had been placed.

"Several years."

"I take it the show goes well. What ages are the children in your audience?"

"Eleven, twelve, thirteen . . . "

"Do you know where 'troldspejlen,' the name of your program, comes from?"

"No. Well, I know it means the magic mirror."

Lucas seemed surprised by her "no." He assumed that all Danes know Andersen the way that all English know . . . well, come to think of it, he wasn't sure what they all do know these days. He explained to Karen, "'The Trolls' Mirror,' of course. It's an Andersen story."

"Tell me."

"It's the lead story in *The Snow Queen*. It seems that a troll, out of sheer perverseness, built a magic mirror that reduced certain things and enlarged others.

"Can't science do that now?"

"Of course it can," Lucas answered. "In mathematics, it's called an inverter. Everything outside a circle is relocated inside the circle and is diminished in size. Everything inside the circle is transported to the outside of the circle and is enlarged."

The waiter brought a carafe of red wine and two glasses.

Lucas went on. "Now, there was more to it than that: The magic mirror was able to invert attributes other than size. Good and beautiful things were reduced in their goodness and beauty, while bad things were enlarged in their badness."

"Couldn't science do that as well?" asked Karen.

"Create a moral attenuator? Hmm. Well, after the troll made the mirror, he had a lot of what he called 'fun' applying it to people.

I suppose he wanted to show the world that appearances are often the reverse of reality. After a while, this fun, like all fun, began to pall. And he got another idea."

"What was it?"

"He would fly the mirror up to heaven and have great fun with the celestial beings. Even in the heavenly realm, he supposed, there was sham that needed to be unmasked. So he flew the mirror up to heaven. When he got there, the mirror suddenly cracked and broke into a million million small slivers of glass. The tiny pieces floated down to earth, and one of them entered the eye of a little boy."

"What happened then?"

"That's the end of the story of the magic mirror. The other stories in *The Snow Queen* relate the adventures of the boy in trying to remove the splinter."

"Did he?"

"Well, you'll have to read the stories for yourself. I think the story of the magic mirror is highly symbolic. Most of Andersen's stories relate to some perception or experience in his own life. The action of the magic mirror symbolizes the discrepancy between what he felt other people thought about him and what he himself felt to be true. The shattering of the mirror says that the two views are ultimately irreconcilable."

"So my television program is pretty deep stuff?"

"Andersen wrote for adults as well as children. If you want to treat things philosophically, everything in the world is deep. All mirrors are magic. Did Thomas Gray come with us to the restaurant? No? Now don't tell her what I shall tell you quite confidentially: Thomas Gray absolutely hates to look into a mirror. On the other hand, years ago, I had a canary. I used to let him out of his cage. He would fly around in the room and perch on the frame of a wall mirror and look at his reflection for minutes at a time. Then he would burst into song.

"The concept of a mirror—any mirror—is profoundly symbolic. Think of Narcissus looking at his reflection. Shakespeare has Hamlet, the Prince of Denmark, say that an actor must hold the mirror to nature. Oscar Wilde created a portrait that grew old and evil even as the person it depicted remained youthful looking and innocent.

"The water's surface, the mirror, the portrait, the camera, the television program, the play, the work of art—these are all things that transform our perceptions of reality, and the mirror is their chief symbol. Is the portrait that of the sitter or, as some have asserted, is it a portrait of the artist? Is it the spectator that art mirrors, rather than life? What does a camera show? Absolute objectivity or the manipulation of the photographer? What will your afternoon spent in shooting me show? Hans Christian Andersen? Lucas Fysst pretending to be Andersen? Your vision as a successful producer?

"And then, if we listen to the Swami, he tries to teach us that the knower and the known are one; the observer and that which is observed are one. For me, Andersen's story of the magic mirror cuts close to the fundamental question of the relationship between art and reality. The shattering of the mirror represents the impossibility of reconciling the subjective and the objective—the impossibility of eliminating the paradoxical feeling we often get of having created a universe that already exists."

The waiter brought in two dishes of pasta.

"When we get through editing," said Karen, "I'll send you a tape. Then you can make up your own mind about the observer and the observed."

Thomas Gray, who was a bit late getting to the Cafe Italiano from Fyrskib Nr. XVII, caught only the last bit of the conversation.

"There's a difference, I should hope, between the pasta and the pasta maker," she declared. "I doubt if the average Hindu in India thinks along the lines of the Swami. There has always been an export trade in ideas."

. . .

The battle was won; the work was done. Thomas Gray planned to return to Cambridge as she had come, on the *King of Scandinavia* to Oslo and thence back home. Lucas still had some things to do before he could leave.

The boat to Oslo was scheduled to depart at 17:30. At 17:15, Thomas stood forward on the upper deck, barely visible from the street level. On the dock, waving goodbye, were Lucas, the Swami, Dr. Ahlborg, Karen, and Lars Jakob Pedersen.

At 17:21, the Queen, accompanied by Mewsli the Kongekat, appeared quite informally and waved goodbye.

"Farvel! Farvel! Vi ses! Bon voyage! Benefit to all, to all!" shouted the seers-off.

"Ta ta!" answered Thomas.

At 17:30, the ship's horn bellowed forth. It seemed to Lucas that the smoke issuing from the stack rose up and formed itself into the shape of the black stovepipe hat that Hans Christian Andersen used to wear, a "huge cloudy symbol of a high romance."

25

A Good Bit More Is Revealed

Most of the Fellows who had been at Pembroke High Table and had gone into the Combination Room for apples and port had now left. Still in the room, in alphabetical order, were Dr. George Apodictou, Lucas Fysst, and Professor Ian Plumbley. It should be pointed out that Apodictou was Reader in Eastern European History and was one of Lucas Fysst's closest friends.

"You've been away from College a long time, Fysst, and I know you have a story to tell us," urged Ian Plumbley. "I can tell by your face."

"Yes, I have a story, and I have no intention of telling it in a short version."

"Well, give us the medium-length version."

Professor Plumbley prepared for the narration by taking an apple and paring it slowly into a long spiral. Dr. Apodictou loaded his pipe and tamped down the tobacco.

By midnight Lucas had told his story essentially as we have had it in this book up to this chapter. He then went further.

"The manuscript that was offered me at the Viking Ship Museum for five thousand pounds was of the well-known Andersen story "The Elf Hill." The story was one of the manuscripts that had been stolen quite recently from the Andersen Museum in Odense. The manuscript didn't interest me at all. I was looking for something genuinely new. But in view of the museum theft, an event that was in the Copenhagen papers for weeks and weeks, it was important to identify what I was looking at.

"I suspected that what was offered was a forgery and that Dr. Kalamos was the forger. Why Dr. Kalamos? Well, for one thing, there are not too many people in the business, and for another, the price was low. Kalamos thinks of himself as an artist, not as a gouger. Did I call Kalamos a forger? Perhaps that word's a bit too strong. Let me be charitable and say he is a simulator.

"Do you know there is a steady market for simulations? For example: illuminated manuscripts. Do you know you can buy a simulated version of Dante's copy of Vergil, complete with interlineations and wormholes?

"Dr. Kalamos is a very interesting fellow. Yes indeed. He's a great scholar and a master craftsman. I've visited his studio in Vienna. He wouldn't allow me into his simulation laboratory, of course. Dr. Kalamos doesn't do much harm in the world. He sells only to the very wealthy and at bargain prices. The game of collecting is a game the rich play. A unique item is worth everything, yet it's worth absolutely nothing.

"The rich set the prices, largely. They know that forgeries abound. It's part of the game. A famous forger once said, and said truly, that there is no false art—only false ascriptions. The sea is full of waves. One has to learn how to sail. What exhilaration is there on placid waters?

"A unique item has to be authenticated by an expert, whose opinion may be matched against that of other experts. What games do the experts themselves play? It varies from person to person. I suspect that whenever Kalamos comes up short in legitimacy, he has plied his craft for the simple thrill of deceiving other experts. I must say, though, that on several occasions he has been tremendously helpful to me."

"What convinced you that you were offered a fake?" asked Plumbley.

"For one thing, the offer came too close to the time of the theft. It may take years, even decades, for an authentic piece to come on the open market after it's been stolen. However, that wasn't really what convinced me. I hypothesized that the manuscript was of Kalamos's manufacture because of its high quality. But hypothesis is not confirmation. Now, I've been in his studio. I know that Dr. Kalamos is inordinately fond of bratislaver, a certain sausage that one finds in Vienna. It's powerful stuff, and Kalamos keeps his atelier in a bit of a shambles. The overtones from the bratislaver sausages enter all his productions very slightly, and Thomas Gray picked up on this immediately."

"By Jove," said Ian Plumbley.

"Splendid," added George Apodictou.

Lucas resumed. "All in all, the adventure in the Viking Ship Museum was an interesting diversion but fundamentally a waste of time.

"Now you may ask the question How did we, Thomas and I, find the unpublished manuscript called 'The Wienerbroed and the Cat'? Well, I shall tell you. It occurred to me, after a certain amount of preliminary spadework, that one of the principal candidates for the location of the manuscript was the Kongelige Bibliotek itself. If an item has been misfiled in a large library, then it is as good as lost. Oh, every once in a while someone may happen upon such an item and call it to the attention of the librarians, but this still leaves thousands of items misfiled. If the library allows readers access to the stacks, then very often a sign is posted for them not to reshelve the books they've taken down and looked at. But the library staff itself makes errors. People are only human, thank God.

"The problem, then, was to find a misfiled item in one of the great libraries of the world whose holdings are in the millions of items—the proverbial needle in the haystack. I had to ask myself, What kind of error might have been made? I could think of many possible kinds, and initial follow-ups on them led nowhere.

"That is, until Thomas Gray pointed the way to me. How she gained insight into the matter is beyond me, but there you are.

"You will recall that I got a letter from the Queen's Secretary inviting us to tea. I read the letter, marked down the date in my black book, and threw out the letter and the envelope, thinking no more of the matter for the time being. The envelope lay there for days and days. Then Thomas gained insight—illumination, if you will. She would not let me forget. She scrounged around in the wastebasket and pulled out the envelope and made sure it would attract my attention. She did this several times, because the first few times, I simply threw the envelope back into the basket.

"Finally, something clicked. I saw that the invitation had been addressed to Fysst Lucas. I remembered also that when I had opened up a little account at Den Danske Bank, the account book and the plastic access card were also made out to Fysst Lucas. I formulated the hypothesis, correct or incorrect, that Danes don't care to distinguish between their first and last names. If one looks in the Copenhagen phone book, one finds page after page of Andersens, Carlsens, Christiansens, Hansens, Jensens, Larsens, Nielsens, Pedersens, Svensens, and so on, and so many with essentially the same first names, that in order to distinguish themselves, people list a middle name or their occupation along with their address.

"I formulated the further hypothesis that when the manuscript —the missing and unpublished manuscript, that is—originally arrived in the Kongelige Bibliotek, the accessions clerk accidentally catalogued the item under one of the six permutations of H.C.A.'s name. I made a list. The variations are as follows:

> Hans Christian Andersen
> Christian Hans Andersen
> Christian Anders Hansen
> Anders Christian Hansen
> Hans Anders Christiansen
> Anders Hans Christiansen

"I went through the old catalogue systematically, looking up the additional five names. (I did not need Dr. Ahlborg's assistance for this.) With the fourth name, I was successful. Pure luck, shall we say. Luck plays a role in all discovery, but it is essential that one know where and how to look. Also how to recognize what one has found after one has stumbled on it. Luck favors the prepared mind."

Thomas Gray walked into the Combination Room unnoticed by the men present. They're talking about me, she thought to herself with unchecked pride; I am the prepared mind.

"Under Anders Christian Hansen I found several items. The second read: *Den Danske Historie*, Koebenhavn, Gyldendal, 1893, 8vo. And there was a notation at the bottom of the card that read (I translate): Supplement in pocket. Danish emigration: Wienerbroed. Ms. 5 pp.

"When I got the book and pulled out the supplement, I knew that The Search was over. 'Eureka!' I shouted aloud in that quiet library. 'Eureka! Eureka! I've found it!' I sang,

'Finita jam sunt proelia

'Now is the victor's triumph won
O let the song of praise be sung!'

"I stood up and danced, holding the manuscript in front of me as my partner. Readers thought me mad. They shushed me. I had indeed found what I was looking for. It was a story called "The Wienerbroed and the Cat." It was written in Danish, naturally. And it was pretty much as I had dreamed it in English. Explain that, if you can. I cannot even now dream in Danish.

"Of course, I proceeded to judge the authenticity of the manuscript, examining paper and ink, comparing handwriting, tonality, stylistic embellishments, and subject matter, all the while alert to anachronisms and neologisms. All this took me a bit of time. Some of my colleagues, who are much more mathematically inclined than I, would submit a manuscript to statistical tests. They would compute average sentence length and relative frequency of nouns, adjectives, and verbs. There is commercial software available for doing this.

"I asked myself, could this be a forgery? If so, by whom and for what purpose? Could the author have forged his own forgery? And if so, would it be proper to consider it as a part of his authentic oeuvre?"

George Apodictou interrupted this monologue. "Didn't you tell us once that you played a part in the unmasking of a literary forgery by—what's his name—Karl Hase of a manuscript on Byzantine history? And wasn't Hase a considerable scholar?"

"Why, yes," answered Lucas Fysst, somewhat perplexed and a bit embarrassed. "And that was a much deeper and more dangerous kind of forgery. Hase's work was a mutilation of history itself. And some day I intend to write about Hase's connection to Napoleon's theorem. That's another lovely piece of detection on Thomas's part."

"I simply wanted to point out," continued George Apodictou, "that common wisdom has it that it takes a thief to catch a thief. I think you dreamt the whole thing up yourself. You planted some kind of manuscript in the Royal Library."

Thomas Gray's whiskers twitched with considerable annoyance as Lucas, red in the face, answered Apodictou's charge.

"My dear Apodictou. All you need do is to go to the Kongelige Bibliotek in Copenhagen and look up Anders Christian Hansen in the old card catalogue. The item is not yet in the computerized catalogue. Then get his book out. Pull the manuscript from the pocket that has been pasted in the back. Notice that the manuscript is stamped with the circular Kongelige Bibliotek stamp and the access date, September 22, 1897. The purple stamp pad ink is now quite faded. Read the story. Notice the author's handwriting and signature. Perform whatever statistical tests you care to on the literary style. Then come back to Cambridge and tell me how all this is possible, without denying that I am an honest man or denying that Andersen, indeed, had a higher regard for cats than was formerly supposed."

Ian Plumbley put his hand over his mouth and yawned quite audibly. "I'm sure that some day, Fysst, you'll tell us the full story," he said. "Now to bed. To bed. It's already tomorrow."

Stories can never be told in full, thought Thomas Gray. The world is a story without an

End.

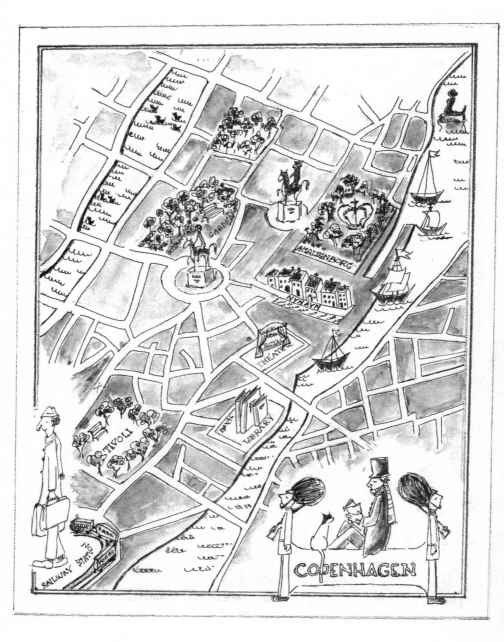